INHERITANCE

INHERITANCE

The Evolutionary Origins of the Modern World

HARVEY WHITEHOUSE

The Belknap Press of Harvard University Press
Cambridge, Massachusetts
2024

Original edition first published in the United Kingdom by
Hutchinson Heinemann, Penguin Random House UK, 2024
First Harvard University Press edition, 2024

Typeset in 12/14.5pt Dante MT Std by Jouve (UK), Milton Keynes

Library of Congress Cataloging-in-Publication Data

Names: Whitehouse, Harvey, author.
Title: Inheritance : the evolutionary origins of the
modern world / Harvey Whitehouse.
Description: First Harvard University Press edition. | Cambridge,
Massachusetts : The Belknap Press of Harvard University Press, 2024. |
"Original edition first published by The Random House Group Limited,
London"—Title page verso. | Includes bibliographical references and
index. | Summary: "Anthropologist Harvey Whitehouse offers a
sweeping account of how three ancient biases—conformism,
religiosity, and tribalism—shaped humanity's past and imperil its
future."—Provided by publisher.
Identifiers: LCCN 2024003421 (print) | LCCN 2024003422 (ebook) |
ISBN 9780674291621 (hardback) | ISBN 9780674298064 (pdf) |
ISBN 9780674298071 (epub)
Subjects: LCSH: Human evolution. | Social evolution. |
Human behavior—Evolution. | Human behavior—Social aspects. |
Heredity, Human. | Religion and culture.
Classification: LCC GN281.4 .W49 2024 (print) | LCC GN281.4 (ebook) |
DDC 303.4—dc23/eng/20240312
LC record available at https://lccn.loc.gov/2024003421
LC ebook record available at https://lccn.loc.gov/2024003422

Dedicated to Merridee, Danny, Sally, Delilah, and Patricia

Contents

Introduction: An Unnatural History of Humanity

As an anthropologist, I am used to people looking at me as if I am odd, or stupid, or both.

I was on a tropical island in the Pacific Ocean that would become my home for two years. This was my first time carrying out fieldwork and I was following people around as they laid out offerings to their ancestors in the cemetery temple. As a PhD student at the University of Cambridge in faraway England, I had come to immerse myself for two years in one of the least studied indigenous cultures of Papua New Guinea. And here was a tribe, deep in the rainforest, whose language had never been written down, whose village had no electricity or running water, and whom few people outside the region had even heard of. I dutifully set out to participate in daily life, to interview people as they went about their activities, and to record everything in notebooks.

The temple itself looked like any other house in the village, made from materials gathered in the surrounding forest using axes and bush knives and with a roof made of grass; however, it fulfilled a very distinctive religious function. As an eager and earnest student, I felt obliged to investigate this idea of feeding the ancestors from every possible angle. *Could ancestors pass directly through solid objects, like the plaited bamboo walls of the temple? Did they* physically *swallow the food offerings left for them? Would the ancestors be pleased by the offerings laid out for them?*

With each question, my friends' faces displayed ever-greater bemusement. Within the first few months of fieldwork, people had come to expect my incessant and naïve stream of questions. On the whole, they were remarkably indulgent, showing seemingly boundless patience in sorting out my misunderstandings. However, this latest line of enquiry took things to new levels of absurdity. *Of course* the ancestors could pass through the walls. *Of course* they didn't physically swallow the food, which would be ridiculous. And *of course* they would be pleased by the offerings; why else would we be making them?

This final answer captured my interest more than the others. 'Aha!' I blurted out. 'So, the ancestors have thoughts even though they don't have bodies?' My informants blinked at me in surprise. *Of course* the ancestors have minds, they said. And perhaps, they were polite enough not to say, we have made a mistake in inviting this idiot to participate in our most sacred rituals.

During my anthropological training, I was encouraged to set aside everything I thought I knew and to approach the process of observation in the field as open-mindedly as possible, without preconceptions and assumptions drawn from my own cultural background (the sin of 'ethnocentrism'). But the truth was, even as I asked my questions about the ancestors, I knew what the answers were going to be. Why? Because my hosts were correct: these *were* the obvious answers. In fact, they would be obvious to anyone, anywhere. You could ask the same question about the dead in virtually any human society on the planet and you would find people ready to say more or less the same thing. Everyone is familiar with the idea that spirits are incorporeal and yet still have minds that allow them to emote, to remember, and to understand what we are saying. Equally common is the idea that our minds and bodies are separable, and that our spirits live on after we die. Think of the ghosts in English country houses. Or the spirits in Afro-Brazilian possession cults. Or the ancestors in China.

Such ideas are universal because they are rooted in human nature, resurfacing with each new generation.[1] These naturally occurring beliefs are among the most distinctive characteristics of our species. Judging by the observable behaviours of chimpanzees, bonobos, and gorillas, our closest primate relatives do not conjure up a world in which spirits of the dead demand to be fed, placated, or petitioned for assistance. And yet in human societies, such ideas proliferate like wildfire. While the specific features of these beliefs are expressed in endlessly different ways in different cultural groups, their basic components crop up time and again through our history.

In other words, the things we transmit from one generation to the next – our cultural traditions – have taken some remarkably diverse forms. But all are ultimately rooted in our evolved psychology. This combination of our culturally evolved traditions and biologically evolved intuitions constitutes our collective inheritance as a species, passed down to us by countless generations of our ancestors. This is a book about that inheritance and the dangers of squandering it. And it is a book about how we might invest our inheritance more wisely in the future.

Human nature and our unnatural history

The most basic building blocks of our common inheritance are three natural biases that have been repeatedly observed in all human societies. The first is conformism: the fact that we avidly copy others, sponging up the rituals and other customs of the groups we are raised in, including laying out food offerings in special buildings if that's what people around us happen to be doing. The second is religiosity: that we are naturally inclined to acquire and spread ideas about gods, spirits, and ancestors. The third is tribalism: our oftentimes passionate loyalty to groups, whether that means laying on lavish feasts or risking life and limb on the battlefield. These three biases are crucial to understanding how and why the history of the world has unfolded in the way it has.

But while I argue these beliefs have natural foundations, this book is not a work of crudely reductionist evolutionary psychology, arguing that human behaviour is somehow genetically determined. It is a book about how our natural biases have been harnessed and extended by thousands of years of cultural evolution, enabling us to overcome the limitations of nature so that we can cooperate in increasingly large-scale societies. Cultural evolution has augmented and extended our natural predispositions and susceptibilities.

Together, these two evolutionary processes – biological and cultural – have produced the cumulative wealth of human knowledge and technology that define the modern world.

The process of harnessing our natural dispositions and overcoming their limitations happened in different ways at different times and places. But there are also striking patterns lurking beneath the diversity. To grasp the deep structure of the process, it helps to think of the inhabited world as a giant garden. The wondrous variety of plants in it are like all the diverse patterns of culture found throughout the world. Many of the grasses and bushes grow wild. We can think of these as cultural practices rooted in our natural biases to conform, to believe, and to belong. Wild plants sprout everywhere, despite all efforts to weed them out or at least bring them under control in the name of science, religious orthodoxy, or some other source of authority.

Nevertheless, there are also plant species in the garden that have been deliberately planted – trees bearing exotic fruits, whose thorny defences against predators have been tamed through selective breeding. This is where our natural dispositions and susceptibilities have been harnessed and extended – cultivated, in other words – to produce a great diversity of cultural systems. Such trees represent the first state formations and organized religions in world history in which autocratic rulers first emerged and humans were often killed to please and placate the gods. Although some species of these fruit trees have long since been felled, they were once widespread across many regions of the garden. And, because their stumps remain scattered around, they are still a haunting presence in the garden thousands of years later.

And finally, there are cloned trees planted in neat rows across the length and breadth of the garden. These are the highly unnatural social systems that define our lives today, the product of thousands of years of cultivation. Some of these trees are surrounded by wild plants, as well as diverse varieties of domesticated vegetables and fruit trees – just as in many parts of the world, folk religions and ancestor cults continue to flourish alongside the most tightly policed doctrinal religions. But in some portions of the garden, all sense of wilderness has been lost. These sections are given over to regimented plantations of conifers, just as highly secular systems of governance and education have crowded out most of the wilder varieties of religiosity. Life in this garden, just as in actual landmasses of the earth, is continually changing over time, as seeds are carried around on the wind and the gardeners adopt ever-changing horticultural strategies.

Now imagine translating the layout of this garden into something resembling the geographical distribution of cultural systems around the world, as they became established and spread. The Middle East, northwest India and Pakistan, and Qinghai Province in China resemble parts of our garden in which all forms of life flourished in the sequence I have described: first came relatively undomesticated beliefs and habits arising from our natural intuitions, next came the first complex forms of social organization and religion, and finally there arose the highly cultivated forms of the modern world. In other regions – including North and South America – some of these features were imported, arriving with, for example, European colonization. But even in places where the full range of plant life first emerged and spread, the earlier ones often continue to persist today – wild plants and cultivars all growing together. For example, ancestor worship remains widely practised across much of East Asia and it is still possible to describe certain leaders as somewhat divinized (e.g., the King of Thailand or the Pope). Once the model for these older forms of religious organization had become established, they seldom completely disappeared from the cultural repertoire. Often, they temporarily faded only to return through revolutions, reformations, insurgencies, and other seemingly cyclical processes of upheaval.

Understanding this interplay of human nature and culture is vital to addressing the problems facing us in the world today. If we are bad gardeners – ignoring the accumulated knowledge of horticulturalists in the past – then we may end up degrading the soil and running out of food both literally and metaphorically. The key to our survival does not only lie in planting more of the seeds we want to grow. It also lies in recognizing the natural limitations of the garden and choosing to work around them rather than against them.

If this approach sounds uncontroversial, in the world of academia it is anything but. Unfortunately, it is a basic dogma for most social scientists that one can and should study social and cultural systems without considering human nature. On the face of it, this is a strange position to adopt. Unless swayed by some form of religious fundamentalism, most educated people today would acknowledge that our brains evolved under natural selection, along with other features of our biological makeup. And yet so many social scientists are quite content to assume that we are blank slates when it comes to the creation and spread of cultural beliefs and practices.[2] This involves ignoring or dismissing much of the evidence from the cognitive and behavioural sciences.[3] To confuse things further, crude forms of socio-biology have become popularly associated with right-wing attitudes, while blank-slate thinking is associated with more left-leaning or liberal attitudes – although it could just as easily have been the other way around.[4]

But why would anyone want to cede all discussion of 'human nature' to the most reactionary and simplistic commentators? And why would anyone want to adhere to a set of doctrines denying or marginalizing the role of human nature in culture and history in the first place? As this book will show, nature and culture – whether we like it or not – are intimately entwined in all aspects of our lives.[5] In the chapters that follow, I will describe many diverse ways in which culture is shaped and constrained by ideas that come naturally. Our natural insights are often quite basic in their origin: for example, all primates develop intuitions early in life about the nature of solid objects – we have some form of 'intuitive physics'.[6] However, only humans make the imaginative next step of postulating beings who defy their intuitive expectations of the physical world – witches, ghosts, ancestors, and other beings capable of passing through solid walls and floating up into the sky.

Natural intuitions feed into our social systems in strange and unexpected ways. To take just one example, our intuitions about supernatural beings are also associated with intuitions about social dominance in ways that are consistent across cultures. My colleagues and I have shown in lab experiments that when babies observe an agent capable of floating around like a ghost or a flying witch, they expect the levitator to win out in a confrontation with a rival who lacks such powers.[7] To put it more pithily, we naturally look up to supernatural beings. This could help to explain not only why stories about superheroes – from Santa to Superman – are so popular with children but also why magical beings and their earthly embodiments are so often venerated in human societies. Human cultures have built on these ideas over the course of history to produce ever more elaborately cultivated visions of the supernatural as ungraspably transcendent – ideas that are in some cases so far removed from our natural intuitions that they require a great deal of learning and expertise to pass on from one generation to the next, from the Holy Trinity to the Noble Truths of Buddhism.

All this is why I consider this book not so much a work of 'natural history' as of 'unnatural history'. Certainly, the biological evolution of human psychology under natural selection is an important part of our inheritance – what makes us who we are today. But the modern world is also a product of processes of cultural evolution – the survival of innovations that spread successfully and the winnowing out of the rest. Human nature has shaped and constrained our unnatural history, but it has not determined it. If we want to understand how the garden of human societies became larger and more complex, we need to understand both our natural intuitions and our cultivated societal norms and institutions. This is not a case of nature versus culture, but of understanding how the effects of nature and culture work together.

When I first started making this argument more than thirty years ago, it was a dangerous thing to do. Anthropologists were supposed to study culture in a kind of hermetically sealed bubble, inaccessible to science. Attempting to explain how culture was shaped and constrained by our evolved psychology was condemned using words like 'reductionist' or 'scientistic' – as if reductive reasoning and scientific method were sins. If I was cheeky enough as a student to suggest otherwise, I would typically be rebuked with the phrase: 'That is not anthropology.' Taken as merely an observation about the state of the discipline, it was true. But it was also a veiled threat of exclusion from the tribe. For someone not yet in a tenured academic post, it seemed like a choice between professional suicide or exile.

Soon, however, I realized there was another way out. By the time I returned from the rainforests of Papua New Guinea, I had learned the knack of living safely within a community without necessarily adopting its belief system as my own. Just as I had learned to do with my friends in the rainforest, I began to wonder whether I could become a participant observer at my university: quietly watching the ways of my fellow academics, all the while developing a distinctly new way of thinking about human behaviour.

A new science of the social

At its heart, this new way of thinking was about bridging the silos of academia. Social scientists – just like any other human groups – are quite tribal. By the time I was in my early thirties and well established as a university lecturer, I had grown accustomed to the subtly different insignia that political scientists, sociologists, archaeologists, and so on, bore to indicate the different tribes of which they were members – and how closely they stuck to them. Every year, at exam time, I would have to take students' exam scores to the different departments on campus. I would find myself confronted in one room with colleagues resembling hippies (mostly sociologists), in another with people in pinstriped suits and skirts (lawyers), in yet another with academics bedecked in body piercings (ethnomusicologists), and in the last with baggy corduroys and leather elbow patches (historians). Not only did they form into tribes with distinctive identities, but each tribe kept itself to itself – except to bark out lists of student marks at each other during exam periods.

The differences I had noticed between groups of colleagues were tell-tale symptoms of the silo problem – that is, the tendency of specialized groups to become isolated from each other. The reason this is a problem is not because people end up dressing differently, but because it impedes the sharing of theories, methods, and findings that would benefit all concerned. The more time I spent talking with experimental psychologists, for example, the more I realized their research questions could be informed by anthropology and the more their methods could be used to address anthropological problems.

Consider an enigma that I spent large portions of my early career grappling with. During the two years I was doing fieldwork in Papua New Guinea, I frequently observed people placing money in special cups in order to be absolved of sin. Those same people donated a large proportion of the money raised in this way to charitable causes. I wanted to know if rituals of confession somehow caused such remarkable acts of generosity. Rituals in which people confess their sins and receive forgiveness are found in a wide range of cultural traditions, including of course the Catholic Church. Do such rituals make us more generous towards those in need, enabling communities living in poverty to survive? The trouble was, we could not tell if that was the case simply by observing people going about their lives or asking them why they do what they do. Observation is not enough. After all, people who confess their sins also do many other things that could just as plausibly explain why they give money to good causes. Maybe they do so to curry favour with God. Maybe they are acting out of feelings of empathy towards those who are suffering. Maybe it has to do with signalling commitment. Even if they are happy to talk about their motivations, it is impossible to know whether this is a realistic account of what makes them do what they do. This is why we need to run experiments.

As soon as I had figured out how to get research funding, I started hiring postdocs trained in the methods of experimental psychology to answer these kinds of anthropological questions. Using one of my first grants, I employed a postdoc called Ryan McKay (who later worked his way up to full professor at the University of London). In one of our earliest studies together, he recruited a sample of Catholics and asked them to recall a sin they had committed in the past and the process of confessing it – and provided them with the opportunity to donate money to the church. But whereas half the sample thought about confessing and being absolved *before* being asked to reach into their pockets, the other half only thought about it *after* the donation task. What we found was that people who had just been thinking about confession were more generous than those who had not.[8] This experiment, partly inspired by the confessing rituals I had studied in Papua New Guinea, offers a simple example of how experimental methods can address the thorniest of questions about why people act the way they do. It illustrates why we need to combine fly-on-the-wall ethnography with controlled experiments if we are to get at the deeper underlying causes of human behaviour.

In the years that followed, my colleagues and I would use this approach time and again: drawing upon ethnographic observations to inspire experiments aimed at explaining the diversity of human social and cultural systems. In the course of this book, I will be describing experiments my colleagues and I have carried out in populations as diverse as football fans in Brazil, revolutionary insurgents in Libya, farmers in Cameroon, religious fundamentalists in Indonesia, martial arts groups in Japan, and chiefdoms in Vanuatu – all the while hopping between academic silos to source the methods needed to answer our questions from different angles.

Long after I had started to use experimental methods to answer the big questions I was interested in, I began to realize that qualitative observations based on field research had another useful application. True, they could help us to generate promising hypotheses that we could then test in experiments, generating new data as a result. But field observations were also data of scientific interest in their own right. To use such data to explain things required some method of comparison. Consider those confession rituals. You couldn't claim that these rituals made followers more charitable based on a single example. But you could reasonably establish support for such a claim if it were possible to show that a very large sample of groups with rituals of confession were more generous than a large sample of otherwise similar groups that lacked this type of ritual. This insight inspired the work of another of my first postdocs, Quentin Atkinson (who also went on to become a full professor, in his case at the University of Auckland). Atkinson and I built a large database of ethnographic observations relating to rituals – over 600 of them from scores of different cultures – which could then be used to test a variety of theories about the causes and consequences of ritual in human societies.[9] The results of this work profoundly shaped the direction of our later research on the power of ritual.[10]

I soon realized that what applies to ethnographic observations also applies to historical records. The conventional way to write a history book is to describe events or features of life in the past in as much detail as possible, based largely on surviving texts and other artefacts. This can provide a rich picture of things that happened long ago and what they meant at the time. However, just like descriptions of things we can observe qualitatively in the world today, it cannot actually explain scientifically the patterns we observe in human history. To do so requires a more quantitative approach. My colleagues and I realized that if only we had enough instances of when various institutions and inventions occurred in different parts of the world, we could establish what preceded what. And based on the logic that causes precede effects we could start to build up a much more detailed picture of the drivers of various changes in human societies over the course of world history. Enter the world of big data and the statistical analysis of past civilizations.

Quantifying history and prehistory – just like combining ethnography and experimental methods – may be regarded as somewhat deviant, a breach of the usual norms of social science and humanities disciplines. Even for those historians and archaeologists who are happy to work within a broad church, there is something discomforting – even distasteful – about distilling features of the past into thousands of little data points. Like ethnographers, historians deal in richly detailed texts, surrounded by walls of footnotes. These texts easily run into volumes of highly descriptive accounts of past events and the social and cultural contexts in which they occurred. But my colleagues and I set out to organize what is known about the past in a very different way by building a massive list of features of social organization, agriculture, technology, warfare, religion, ritual, and so on. And for each item on the list, we have sought to establish whether or not it was present in a large sample of the societies that existed over thousands of years of world history. With the help of scores of historians, classicists, archaeologists, and anthropologists, we have ended up with a vast database of information on the human past that can be analysed statistically – enabling us to identify the factors that helped to drive the rise of ever more complex social systems from the first farming societies to the largest-scale states and empires of the modern world. As we will see, this is a process that has produced some rather shocking discoveries.

As the ambition of this project grew, it came to incorporate an ever-greater array of methods, disciplines, and approaches. The ideas in this book have taken four decades to develop and have led me into long-lasting collaborations with evolutionary theorists, data scientists, and statisticians, while also sucking me deeper into the worlds of historians, archaeologists, ethnographers, and psychologists. It also led me to establish the Institute for Cognition and Culture at Queen's University Belfast and the Centre for the Study of Social Cohesion at the University of Oxford. Along the way, I have led field research in regions as diverse as Asia, Africa, South America, and Australasia, as well as my original stomping ground of Melanesia. I have become deeply involved in work at the world's most important archaeological sites, brain-scanning facilities, and child psychology labs, and been invited into the heart of some of the world's most embittered conflicts and extremist groups. In all of this, I have been trying to carve out a more encompassing approach from the most established shibboleths of these disciplines: a transdisciplinary science of the social.[11] It is this new, scientifically grounded study of our collective past, present, and future that led me to write this book.

The three biases

In the pages that follow, I will introduce the evolved biases that have, over thousands of years, been extended by our forebears and helped to forge the peculiar world we now inhabit. First, I explore the most essential – and idiosyncratic – feature of human nature: our tendency to copy one another even when the purpose of the behaviour is obscure. This compulsion to imitate, I will show, enabled humanity to store the discoveries of earlier generations, allowing cultural traditions and knowledge to accumulate over time. I refer to this aspect of human nature as conformism.

Conformism has always been a mixed blessing. It explains many of our most astonishing cooperative accomplishments and also some of the most appalling atrocities committed by our species. For better or worse, however, a sharp increase in the frequency of ritualized acts of conformism played a key role in the rise of the first large-scale societies. It enabled shared regional cultures to coalesce and new forms of political unification to take shape. It also enabled us to become more future-minded than ever before.

Unfortunately, cohesion and future-mindedness are declining in many regions of the world today with the spread of polarizing politics and unsustainable forms of overconsumption. However, I will show that the same conformist instincts that led us down this path to destruction can also lift us out of it.

The second major bias considered in this book – religiosity – is no less significant to the history of humanity. Religion is not a merely optional addition to the cultural repertoire, as many atheists assume. Still less was it given to us by God, as many adherents to organized religions insist. Rather, I will argue that it is an inescapable by-product of the way our brains evolved. Over the millennia, religion has played a key role in enabling new forms of leadership, facilitating ever more complex social systems. This process took humanity through forms of ancestor worship, systems of inherited authority, divine kingship, and eventually led to the great moralizing religions that have spread to much of the world we live in today.

But as the grip of organized religions is loosening, new forms of religiosity are taking over, largely below the radar. The advertising industry, social media, and news conglomerates are harnessing our natural religious intuitions in ever more creative ways, for example by using such techniques as brand anthropomorphism and behaviour modification. In this manner, small elites are increasingly profiting from our credulity on a scale far greater than any of the organized religions of history. Worse, the costs of this are often borne by those least able to pay. Here again, however, I will show that there are means by which we can take control of the very same features of human nature that global capitalism exploits so corrosively and use them to achieve more consensual outcomes, for example by reforming social media platforms and news outlets. Since intuitive forms of religiosity come naturally to us, we cannot eliminate this aspect of human nature, but we can at least manage and harness it so that it benefits human flourishing and limits its more irrational and destructive manifestations.

Finally, we will turn to the third bias: tribalism. I will argue that while tribalism has produced some of the cruellest acts in history, it has also motivated many peaceful and creative forms of cooperation. And I will explore what we have learned about the psychological roots of group bonding and its effects on our behaviour, for good or ill. Based on this psychological research we will see how our tribalism has been harnessed and extended over the course of history. Some of our findings on this topic have been unsettling. For example, it turns out that one of the most powerful factors driving the rise and spread of civilizations in world history is warfare.

At the same time, these once-helpful tribal urges now pose an existential threat to our species. Warfare can no longer benefit the winners as it has done in the past because military technologies have become far more destructive than ever before. However, I will also argue that the solution to many of our modern-day problems may be more tribalism, rather than less. I will show how the most harmful ways in which our tribal instincts are currently being deployed could be turned to our collective advantage in a host of practical ways, at the level not only of local communities, states, and international diplomacy but through the development of new forms of global tribal identity.

In short, the very same features of human nature that are bringing us down could in theory also be harnessed to reform our economies, conserve our planet's resources, expand our cooperative capacities, and manage conflict more effectively – in other words, to create a new basis on which human civilization can realistically flourish. This is why each part of the book rotates through the three biases – conformism, religiosity, tribalism – in different contexts, but the same order. By circling three times through the three biases that have made us who we are today, we can see both the timelessness of certain features of human behaviour – and the potential we now have to plot a new course.

Humanity may yet destroy the planet via a diversity of methods, including climate breakdown, famine, disease, and nuclear holocaust. Our ability to stave off apocalyptic outcomes will depend on whether we can harness our evolved psychology in wise and consensual ways: wise in that they are grounded in science; consensual in that they are predicated on shared moral concerns, collective interests, and wide consultation.

Yet if we are able to do so, the future of the world could be better than currently seems possible. And that is why the message of this book is ultimately a hopeful one. Recent breakthroughs in the scientific study of human nature and the evolution of our societies and cultural systems can be applied to all the most pressing problems facing the world today. The purpose of this book is not only to chart the dangers but, much more importantly, to sketch out a road map for peaceful cooperation and sustainable prosperity. The challenge will be to make the best use of our natural and cultural inheritance, rather than to squander it.

PART ONE

Nature Evolved

I.
Copycat Culture

At first, the four-year-old girl seemed hesitant. But after seeing the lady's smile of approval, she forgot about her shyness and reached out for the object with evident curiosity. This girl was just one of scores of participants in an experiment at the University of Texas at Austin, which my colleagues and I had set up to explore how and why young children copy behaviour, even when it has no obvious purpose. Laid out on a table in the child psychology lab were a variety of brightly coloured objects. There was an orange sphere, a blue cube, a purple chess piece, a multicoloured peg board, and – most intriguingly of all – a silver painted box.

In our experiment, children were first shown this odd assortment of objects being used in a short video. In the video, each of the objects was set out neatly in a line on a desk, behind which sat a young woman. The woman lifted each object in turn and performed an action sequence in which it was twirled around or tapped on one or other of the other objects, sometimes repeatedly, before being placed carefully back on the spot where it had originally been positioned. After all the objects had been handled and replaced, the video ended, and the objects were pushed back in front of the child. The big question was what the girl would do next. Would she continue playing with the objects? Would she lose interest and ask if we had any better toys to play with? Would she want to leave? In fact, she did none of those things. Like most of the other children in this experiment, she did something much more puzzling. She *imitated* the apparently pointless behaviour she had just observed in the video.[1]

Buried in this behaviour lies a crucial clue to why humans have customs, rituals, and fashions of all kinds. At some point in childhood, we inevitably come to accept that in our particular community, you greet people by shaking hands or pressing your lips to another person's cheeks, or you have to get your hair cut and decorate your body with special objects, lotions, and clothing, you eat your food a certain way, and so on and so forth. This all soon becomes so obvious that we never question any of it. But for each of us, there was once a point in our lives when we first observed with puzzled fascination that people around us were grasping each other's hands or hanging sparkly objects around their necks. Although we all eventually come to realize that our own bodily adornments and modes of greeting are peculiar and people in other communities behave differently, very few of us ask the obvious question: why do we do this kind of thing at all?

The Austin experiment was our first attempt to answer that question. In the preceding years, psychologists had begun to explore the science of imitation. But many were focusing on a strangely narrow answer. They argued that young children copy seemingly opaque behaviours because they trust older and wiser individuals not to teach them useless habits. That is, even though the weird behaviours of adults might look pointless, we assume that one day we will finally realize what they were for and be glad we took the time to learn them. If that never happens, then presumably we end up forgetting why we copied the behaviour in question, and it is eventually perpetuated simply as a habit or custom.

But as an anthropologist, I had a different idea. Having spent much of my life studying diverse rituals in human societies, I realized that much of the behaviour we copy from other people in our community is never expected to be of any material use at all. Its value lies in what it tells us and others about the groups to which we belong. At a university on the other side of the Atlantic from Oxford, I found a team of psychologists led by Cristine Legare who shared this hunch and were willing to work with me to put it to the test.

Our experiment with that little girl was designed to examine why, from an early age, humans copy behaviour that has no apparent purpose. We wanted to see what happened in two subtly different experimental conditions. Fifty-seven children aged four to five took part in the study. Roughly half of them observed the objects ending up where they started (no 'end goal'). For the other half, instead of all the objects ending up exactly where they began, one of the objects was ultimately placed in the silver box (a very salient end goal, for most children). When we were designing this study, we already knew in advance that preschoolers find boxes intriguing and they *love* to put things into them and take things out. Experience has shown that young children are even more strongly motivated to do this if the box is bright and shiny, like the one in our experiment. And we made an intriguing discovery. Children copied more accurately and introduced fewer innovations and deviations when they watched the video in which the objects ended up where they started – the one without any apparent end goal whatsoever.

If the purpose of imitating behaviour was to learn a technically useful skill, as many psychologists were assuming, then we would have expected modelled behaviour with an obvious outcome – placing an object into a box – to merit more fastidious copying than seemingly pointless behaviour that leads nowhere, ending up exactly where we began. After all, technically useful skills tend to have some noticeable, material impact on the world. In fact, as we shall see, this prediction would be quite reasonable for any other species than our own.

But from an early age, humans do the very opposite. They stick more closely to the script when the behaviour they are copying has no material outcome. Why? Because, I argue, humans are a ritual animal.[2] We are naturally predisposed to sponge up the behaviour of others even when it has no obvious purpose. Humans everywhere engage in seemingly unnecessary behaviours that are unique to their own cultural groups. For example, every human group described by anthropologists or historians has some distinctive set of rules for how to mark major transitions in life (such as birth, marriage, and death), how to celebrate communal achievements, how to memorialize significant events in the past, or how to exert a magical influence over nature. Some of these rules are enacted in solitude, some in families, some in local communities, and some in massive gatherings of thousands of people and more.

And yet if we stopped doing all those rituals tomorrow, babies would still be born, people would still fall in love and move in together, and all of us would one day die. So, why bother with all the rituals? What is it about humans that makes us imitate seemingly pointless behaviours and to sponge up the customs of the communities in which we are raised? Our research suggests that it stems from the motivation to conform – to establish ourselves safely and securely in the heart of the group by copying those around us.

The ritual animal

Since there is no limit to the ways in which locally distinct customs can evolve, human cultures are breathtakingly diverse. In 2017, I was invited by the BBC to serve as chief consultant on the world's first television series devoted to the topic of human rituals and, together, we embarked on the task of creating a documentary series that would span the behavioural diversity of our species around the planet.[3] In one episode, a young woman, Bhumi, was seated on a stage before a large audience on the outskirts of a Rajasthani town in northern India. Her flowing white robes contrasted with her dark, luxuriant hair. The crowd watched with a mixture of empathic fascination, horror, and admiration as each hair on her head was painfully torn out by hand until she was completely bald.[4] Bhumi had decided to give up worldly desires in pursuit of an ascetic existence as a Jain nun. Having her hair torn out was an act of self-sacrifice as she willingly forsook her youthful beauty as an expression of commitment to something of seemingly much greater value.

A moment's reflection tells us that this was not simply an instrumental action of removing unwanted hair from Bhumi's head or causing her to suffer. For a start, why the very specific rules for this act of self-sacrifice? There are seemingly countless ways in which beauty can be defaced and pain inflicted. But even an uninformed observer would appreciate that this is a distinctive local custom carried out in this *particular* way by those wishing to become nuns within this *particular* ascetic tradition. One of the cues to this particularity of the rules is the fact that Bhumi is not alone on the stage but is one of many young women attired and seated in precisely the same way and being subjected to precisely the same treatment.

Bhumi's ordeal was only one of many portrayed in the *Extraordinary Rituals* television series. The documentaries covered a great diversity of challenging rituals from around the world, including bareback horse-racing in Siena, body-piercing in Malaysia, New Age pilgrimages to Nevada, courtship rituals in China, dancing with corpses in Indonesia, Kayapo birthing rites in Brazil, and many more. We may easily recognize the social significance of all these rituals as creating new life, acting out violent conflict for the pleasure of the crowd, giving vent to inter-community rivalries, match-making and marriage, disposing of the dead, welcoming new members of the group, and so on. These are of course universal human experiences. But the ways in which we manage those experiences have given rise to a wondrous variety of ritual forms. Part of the reason why our BBC television series was so compelling to many viewers was that most of the rituals featured would have been new to them and therefore somewhat 'extraordinary', as the title of the series suggested. But, at the same time, the basic psychology underlying all these rituals is much the same everywhere.

At the heart of that psychology is the assumption that the rules for conducting rituals are not constrained by normal principles of cause and effect. I describe this kind of behaviour as irremediably 'causally opaque': there is no known (or even knowable) causal link between the action and some intended outcome. Think for a moment about the seemingly simple act of locking a door. If you accomplish this by turning a key, you may assume that, somehow, the shape of that key is distinctively designed to engage with the barrel of the lock so that this key, and only keys cut in exactly the same way, will cause the bolt to be driven into the doorframe to prevent the door from opening. Exactly what it is about the shape of this key that makes it work, while other keys do not, may be unclear. That is, you might not have a fully specifiable overview of how the particular barrel of your lock relates to the shape of the teeth on your key. Nevertheless, you assume that a locksmith or somebody who knows how to pick locks would understand those details. This type of assumption is the hallmark of what I call 'instrumental' reasoning. It means that whether or not the causal role of an action is transparent, we take it for granted that somebody, somewhere could explain it in full. As such, any puzzling elements are potentially resolvable.

In the case of ritual actions, however, this assumption goes out the window.[5] If we imagine that the turning of a key causes a door to become unlocked and openable – but only in a way that is impossible for anybody anywhere to explain in causal terms – then it becomes a magical ritual. The causal link between the action (key turning) and its supposed outcome (unlocking the door) is irresolvable – that is, nobody could explain the mechanics of why the turning of the key is necessary for the door to be opened. In such a case, we would have to describe the door-unlocking procedure as *magic*. Note that the behaviour of turning a key is identical whether you think that the teeth engage the barrel of the lock in an explicable fashion, or you think that simply turning the key magically causes the lock to open. The important thing is how we make sense of the process.

Not all rituals are magical. Some rituals do not have any specifiable outcome that is accomplished by magic. Think of the Catholic practice of self-crossing, for example. Worshippers do this when they enter church but in most cases they could not say why. They just do it because, well, everyone does it. It's what you do. Whereas unlocking the door to the church is necessary to gain access, nobody thinks the same logic applies to self-crossing. And yet if you were entering a locked church, you would do the motions with the key in the door and also the motions with the self-crossing with equal regularity and predictability. Exactly how turning the key opens the door may be hidden in the mechanisms of the lock but you assume that the puzzle of its inner workings could be resolved. The opacity of self-crossing, however, is irresolvable. Whenever actions are assumed to have an unknowable causal rationale in this way, we know we are thinking ritualistically.

Ritual is all in the mind

One of the consequences of this is rather disconcerting. It means that there actually isn't any such thing as a ritual objectively 'out there' in the world, in the sense of a distinctive type of action that can be called by that name. The concept of a ritual is purely in your own mind – in the assumptions you make about whether or not the action is guided by a knowable causal process. If you assume, however implicitly, that an action is irremediably opaque, then – for you at least – it is a ritual. For somebody else, the same action might not be.

To convey this point concretely, it helps to tell the story of Sylvia's recipe, which psychologists sometimes use to illustrate how humans imitate each other without necessarily understanding why.[6] Sylvia enjoyed cooking and one of her favourite recipes involved roasting a joint using a method learned at her mother's knee. When Sylvia was a child, her mother would always cut off the ends of the joint before placing it in the oven. One day, when Sylvia was a grown woman, her mother came to stay and during this visit Sylvia prepared a joint for the oven in the time-honoured fashion. Her mother remarked upon this in surprise. It turned out that the reason Sylvia's mother had cut off the ends of the joint all those years ago was because in those days the family only had a rather small baking dish. The reason for the behaviour had been purely instrumental.

It is possible that even as a child Sylvia guessed that there was some perfectly practical explanation of that kind, even if it wasn't obvious. In this case, she would have been adopting what I call the 'instrumental stance': assuming that there is some practical end goal to which the behaviour is leading. But it is also possible that Sylvia simply assumed that cutting off the ends of the joint was a convention that had no knowable rationale. If so, she would have been adopting what I call a 'ritual stance' – by assuming that the behaviour in question does not follow any understandable logic of cause and effect. If that is how we are viewing a behaviour and somebody pops our bubble by pointing out the original rationale of the behaviour, then it would immediately cease to be a ritual and become just another technical action.

Now, something rather remarkable happens when you adopt a ritual stance. It means that the behaviour in question – because it is assumed to be arbitrary – could mean potentially anything you like, or anything you happen to be told by someone you trust. For example, in a hunter-gatherer culture, cutting off the ends of the joint might mean that you intend to release the spirit of the animal back into the wild so that it will enter the body of another species of prey that the family will later be able to hunt successfully. In another culture, cutting off the ends of the meat may be something that people from Sylvia's mother's ethnic group have 'always' done – it's part of their tradition and what marks them off from other ethnic groups from the same region.

All these ideas about what makes something seem to us like a ritual behaviour may sound quite complicated. But actually we apply these principles all the time, without even noticing. In fact, that's partly why the story about Sylvia immediately makes sense. We intuitively know the difference between a purely instrumental explanation (cut the ends of the meat so it fits in the pan) versus a purely ritual one (cut off the ends as an expression of your cultural identity). Viewed in these terms, it becomes clear why we live in a world of rituals – and are endlessly adopting the 'ritual stance' in making sense of the behaviours we witness.

Sometimes it isn't obvious whether to interpret something as a ritual or not. Consider drawing the curtains. If asked, most people would probably say that drawing the curtains in their living rooms is a purely instrumental behaviour – for example, city dwellers do it after dark to provide privacy from the prying eyes of passers-by. Or we might say we want to shield the television screen from the glaring reflection of sunlight during a crucial football match. But often we just draw the curtains because it's a daily habit, rather than because we want to accomplish some specific goal. For this reason, if a family from a different cultural group moved in next door and failed to draw their curtains at night or kept them closed all through the day, their neighbours might readily attribute this to differing cultural norms regarding the use of curtains and the social signals this can convey. Or imagine if someone were to tie the curtains together instead of sweeping them apart and using curtain ties – then we would become abruptly aware of a breach of the norms of window-dressing. But in the absence of some jarring departure from the usual script, when a behaviour as routine as 'drawing the curtains' is part of the fabric of everyday life, we are less likely to notice how ritualistic it has become for us.

Spotting rituals, then, is really about catching ourselves and others in the act of adopting a ritual stance: noticing when we have ceased to apply the standard logic of cause and effect to our actions. And yet one of the effects of growing up in a community is that we naturally acquire lots of ritual behaviours without being aware of it. I remember visiting a greasy spoon café in New York for the first time and wondering what a 'hash brown' was. The waitress looked at me as if I were deranged, rather than merely English, and replied: 'Hash browns is hash browns.' They are what they are. But this is only true if you grew up with hash browns. In much the same way, our ritualistic nature often escapes our attention, hiding in open view.

This is one reason why anthropologists often choose to study cultures that are remote from the ones they were raised in.[7] To spot the rituals we are looking for, we need to seek out groups with conventions that are new to us so that the phenomenon we are interested in stands out more starkly, demanding an explanation. This is difficult to do vicariously, by reading books or watching television documentaries. A more effective way of discovering the common underlying features of ritual is to travel to many different cultures and observe behaviour on the hoof, by participating in the lives of local people as one learns about an initially unfamiliar cultural system. Anthropologists often do this by immersing themselves for long periods in a cultural group. In the rainforest of Papua New Guinea, where my anthropological career began, encountering unfamiliar customs made me very conscious of the distinction between ritual and instrumental stances. But by the same token, coming home from long periods of fieldwork also brings the rituals of one's own community into sharp relief.

After two years living in a New Guinea community deep in the rainforest, I had to return to England and participate once again in the now seemingly surreal dining habits of Cambridge colleges and family Christmases. Why did the process of eating at high table need to be preceded by standing solemnly to listen to a formulaic 'grace' uttered in a language few of us could speak, praising a Father, Son, and Holy Spirit who were not physically present? Why place gifts under a special tree and wrap them in coloured paper? Why should one shake hands with an uncle but kiss an aunt?

I had become a professional ritual 'spotter'. Just as some naturalists go in search of birds and butterflies,[8] I found myself hunting for evidence of the ritual stance behind human behaviour everywhere. But in the same way that many ornithologists and lepidopterists eventually become interested in unravelling the evolutionary origins of the traits they are studying, some ritual spotters, like me, have become preoccupied with explaining how ritualistic behaviour evolved in the first place.

The fact that humans routinely engage in ritualistic behaviour presents an evolutionary puzzle. Many species are quite good at learning from observation how to copy behaviours that produce desirable results.[9] Some birds (e.g. certain species of crow) are astonishingly good at it, being able to memorize the exact behavioural sequences required to retrieve a food reward.[10] But even our closest primate relatives wouldn't waste their time copying behaviours that have no obvious practical outcome.[11] Humans, however, do it all the time. Psychologists refer to this as 'overimitation': the copying of behaviour that doesn't contribute to an end goal.[12]

The origins of overimitation

Many studies of overimitation use puzzle boxes – see-through contraptions containing a desirable object that can only be obtained by performing certain procedures in sequence, such as unlocking or opening a series of obstructions in order to retrieve an object.[13] These boxes can be used to see which sorts of behaviour children copy, and which they don't. Studies of overimitation usually involve a single adult going through a complicated procedure before opening the puzzle box – including weird gestures (such as hand-waving) and obviously superfluous actions (such as tapping the box with a feather) – before giving the same task to the child. After observing the model performing these sorts of actions, children will typically copy the useless bits as well as the instrumentally sensible ones. Even more bizarrely, it doesn't seem to make any difference if the experimenter explicitly points out that the model will be performing some 'silly' behaviours that have no relevance to opening the box and explicitly advises the children not to bother copying those bits. The children still copy the unnecessary actions regardless.[14]

What explains this tendency to overimitate? A possible evolutionary explanation is that it allowed our ancestors to pass on useful discoveries efficiently, based on trust rather than exclusively on shared reasoning and evidence. This would have been quite a revolutionary adaptation. Consider, for example, the opportunities this opened up for the discovery of various methods of healing. Many plants have medicinal properties but figuring out which ones have which health-giving benefits is laborious and slow. Individual learning based on trial and error can only get you so far. But if you could trust experts to show you which ones work best, then you could save a lot of time. Most importantly, this would provide a way in which the accumulated discoveries of many previous generations could be passed on without having to start from scratch each time. What works for medical expertise also works for many other domains of cultural learning – from the production of increasingly specialized tools to the construction of dwellings or the tracking of animals. For these skill sets to grow and get passed down through the generations, humans needed to be able to learn new skills on trust.

But how do you know *whom* to trust? When it comes to learning instrumentally useful techniques, a good strategy is to preferentially copy people who are more competent and skilful. That usually means people who are older, wiser, more experienced, have better track records, are relied upon by others, seem confident, get results, and so on. These are the qualities we might associate with *prestigious* individuals, and so the propensity to preferentially copy their behaviour is usually described as 'prestige bias'.[15] From an evolutionary perspective, much of the overimitation observed by psychologists using puzzle boxes and child participants is motivated by the desire to learn something useful from a prestigious expert – typically an adult – who can be trusted not to teach us things that are harmful or pointless. And intriguingly, this trust seems to be invested more in deeds than words. When the prestigious individual tells us not to copy the silly bits, we seem unable to help ourselves and slavishly copy those bits anyway. This would suggest that the bias to overimitate runs deep in our psychology and isn't something we can override using language and explicit reasoning. Some psychologists refer to it as a kind of learning algorithm along the lines of 'copy all, correct later'.[16] That is, when learning at the knee of an established expert, it makes sense to do whatever they do, as a basic starting point, and only gradually winnow out the superfluous behaviours when we are sure that they have no useful contribution to make.

A slightly more nuanced version of this algorithm would be to copy all actions that look *deliberate* (whether silly or not) but not to bother copying obviously superfluous behaviours that are unlikely to be intended to help us learn something useful (e.g., nose-scratching or heavy breathing). There is some experimental evidence that even babes in arms are sensitive to this distinction. In one study, for example, fourteen-month-old infants are shown a video in which a demonstrator turns on a light switch on the table using her head instead of her hands. Infants are assigned to one of two conditions – that is, they are split into two groups, each experiencing a different version of the demonstration. In one condition, the demonstrator is wrapped in a blanket swaddled around her arms, thus making the use of her hands very difficult. In the other condition, the demonstrator's hands are free. When infants observe the demonstrator in the hands-free condition, they are more likely to copy the weird method of using their heads to operate the light switch. Presumably, when the adult's arms are swaddled, the children conclude she had no choice but to use her head to turn on the light – her behaviour is not an active choice, and so is not worth copying. But if the adult with her hands free is doing the same, that means it was a *deliberate* action – and so is worth copying.[17]

One explanation for the evolution of overimitation, then, is that it provides a uniquely efficient way of passing on useful skills from proficient experts to would-be learners. But I have long wondered whether there is a second explanation for overimitation – that we imitate others not only to learn useful skills, but to be part of a group. After all, people routinely copy behaviour that has no instrumental value whatsoever, simply out of a desire to belong. An example is the wearing of ties – a procedure that involves the mastery of knot construction and the careful placement of the knot over the top button of a shirt. Children may spend months trying to learn the intricate and highly specific technique for tying ties – for some (alas, I was one of them) it could take years to finesse this skill to the satisfaction of adults with a critical eye. And yet, let's face it, there's no instrumental reason for wearing a tie. It is too narrow to keep you warm like a scarf and since there is a button at the top of the shirt, it isn't even needed to hold the shirt front together. It's a pointless garment. But then again, that is precisely the point of it. In this case, we are adopting the ritual stance. We are copying one another not to acquire skills from prestigious individuals, but to affiliate with others in the group.

This is not a case of prestige bias but something else altogether – psychologists refer to it as *conformism* bias. And it seems to be one of the most powerful forces driving human behaviour. But it too presents an evolutionary enigma. If humans do indeed copy each other's behaviour to conform – as well as to simply learn new skills – that raises a simple question: what is the adaptive benefit? Conformism could easily get in the way of learning technically useful skills – in fact, it often does. If we do what other people do solely because we want to blend in, then we end up copying all kinds of instrumentally useless behaviours. Often this gives rise to fashions that years later we will look back on with amazement (I still cringe a little when I see photos of myself with rolled-up jacket sleeves and turned-up collars in the 1980s).

The American philosopher Erik Hoffer once observed: 'When people are free to do as they please, they usually imitate each other.'[18] But the evolutionary explanation for this remains unclear. It is quite possible that imitation rooted in learning new skills has more ancient origins than imitation rooted in conformism. Maybe we initially became copiers to help us to use newer and better tools but then eventually this led to the emergence of more ritualistic cultural practices. But whatever the reason, once people started to copy each other's behaviour in order to affiliate rather than learn useful skills, the stage was set for the proliferation of distinct cultural groups – communities defining themselves by their unique customs, beliefs, and practices. There would be no point learning those unique ways of life unless you wanted to be part of the group that uniquely exhibited them and so rituals became the basis for new group identities, opening the way for new kinds of 'us and them' thinking.

If this is the case, then ritual is about more than just learning new skills – it is about forging identities and cleaving to groups. It was an attempt to delve deeper into this idea that took me to that lab in Austin, Texas to explore the mechanics of conformism in early childhood.

The mechanics of conformism

I have been arguing in this chapter that we copy other people for two very different reasons. One is because we want to acquire a useful skill that will help us complete various tasks more efficiently or effectively. For instrumental imitation of that kind, we should be on the lookout for competent individuals as suitable models: we are guided by *prestige bias*. The other main reason for copying people is that we want to be like them so that we can blend into a group. For imitation of that kind, we would be most interested in observing what other people in the group are doing, however prestigious: we are guided by *conformism bias*. But which of these biases is driving which types of imitation – and how might we tell?

In some cases, it should be relatively easy to work out whether imitation is being driven by a desire to learn a technical skill or simply to fit in. For example, if the behaviour in question has a very appealing end goal that we would wish to emulate – and if the model demonstrates skill in accomplishing it via a series of causally transparent steps – then we are more likely to adopt an instrumental stance motivated by prestige bias. Many people of my age in the UK remember the children's television programme *Blue Peter*, whose presenters were continually demonstrating how to make allegedly useful objects out of everyday items, which we (the child viewers) would then seek to emulate at home, usually to the annoyance of other family members. My poor mother in particular was continually vexed by the mysterious disappearance of milk bottle tops and bottles of washing-up liquid requisitioned for the creation of yet another pencil holder or useless ornament.

But some behaviours demonstrated by trustworthy models are more ambiguous. Sylvia's recipe, discussed above, is a good example.

Was her mother observing a traditional custom of her tribe by cutting off the ends of the joint or was she simply trimming the meat to fit the pan? And which interpretation drove Sylvia to copy her mother? This may seem like an unusual scenario, but the process of learning how to become a member of the group presents seemingly endless challenges of this kind.

It is easy to forget how difficult it was to learn the conventions of our cultural group when they were new to us. I can remember a few examples from my own childhood. A case in point was my first term at infants' school. At the end of each day, the children were required to place the chairs on the tables. I later realized that this was to help the cleaners sweep up once the building was empty. But when I was first getting used to this novel custom, the most salient feature of it was that everyone in my class did it at once, often rather suddenly when the school day was at a close. Since most of us were eager to blend in, we participated enthusiastically. One day, in my eagerness to conform to this collective ritual, I pulled away one of the chairs at the precise moment that the teacher was attempting to sit down on it. Any physical damage this inflicted (I still remember her grimacing in pain) was surely exceeded in intensity by my feelings of shame, which endured long after – and which is dimly evoked by every social *faux pas* I have inadvertently committed since. The dominant emotion surrounding this episode was not empathy and sorrow for having inflicted pain, or even fear of punishment for having injured a figure of authority. Far more harrowingly, it was an intense dread of ostracism from the tribe for having failed to perform correctly one of the most important rituals of the day. My apprehension about a scolding was overshadowed by a much more potent terror of ridicule and rejection by my peers.

In this case, the copying of others' behaviour seems motivated not by the desire to learn, but the desire to conform. And if my experience was anything to go by, failing to adequately perform the ritual was altogether scarier than failing to learn an instrumental skill. This fear sets in early, and never leaves us. With that in mind, I have long wondered whether the bias to conform – rooted in our fear of rejection by the group – is the strongest driver of overimitation among children; stronger even than the need to learn new skills.

But how might we test that hypothesis? Well, if the key motivation for undertaking rituals is a desire to conform, we would expect children to be more likely to enact them when there are strong social cues around them indicating that others in the group are doing likewise. If we only ever see one person tying a knot in a long thin strip of cloth and arranging it under the collar of their shirt, then we will probably think they are very odd and would be unlikely to imitate the behaviour. But if every child in one's school is doing the same thing, it's a very different matter. In other words, seeing multiple individuals behaving in ways that would otherwise seem crazy should serve as a strong cue that we should follow suit.

To explore this idea, we devised a series of experiments in which actors in a video carried out an unfamiliar behaviour using a peg board with differently coloured levers on the side – pressing any one of the levers down caused pegs of the corresponding colour to pop up.[19] The actors in the video would press the levers in a distinctive sequence. There were four versions of this performance, however, and every child participating in the experiment got to view one (and only one) of them. Each of the scenarios modelled a different kind of social cue, relevant to conformism.

In one video, two adult models carried out the strange sequence of actions with the peg board in precise synchrony once only. This made it very clear to the child observer that the sequencing of the action was being done quite deliberately in exactly the same way by two grown-ups, suggesting that it was rule-governed behaviour. The element of synchrony was included because it was well established that moving in synch is a very widespread feature of collective rituals cross-culturally: think of group dancing, chanting, choral singing, marching, parading, or drumming. In another version of this scenario, the child got to see the two models carrying out the behaviour synchronously two times over – a sort of 'double whammy' of inducement to imitate. In a slightly 'weaker' version, the child saw two different models perform the action in succession, but not in synch. And finally, in the weakest condition, the child saw a lone model perform the action twice over, emphasizing deliberateness but diluting any impression that the actions could be seen as conforming to a shared script.

In addition to being exposed to only one of these four kinds of videos, each child received one of two kinds of verbal cues: an instrumental, goal-oriented framing or a conventional, ritualistic one. To produce an instrumental framing, the experimenter exclaimed: 'She gets pegs up! Let's watch very carefully. She gets pegs up.' The focus, then, was on the intended outcome: getting the pegs up. For the ritual framing, by contrast, she exclaimed: 'She always does it this way! Let's watch very carefully. She always does it this way.' This time, the focus was on conformism – the idea that this is how it should always be done. Following this setup, each child was given the peg board to play with but no explicit instruction to imitate. Most children did, of course, copy much of the behaviour they had observed in whichever version of the video they had seen. The question was, which of these conditions would produce the most accurate copying behaviours?

The results were fascinating. They showed that children were more likely to reproduce exactly the peg board behaviours modelled to them after receiving the ritualistic verbal cue – that is, the cues that placed an emphasis on the need to conform over the need to learn a skill. The children who received these conformist cues were also more likely to describe the behaviour afterwards as something socially prescribed (e.g., 'I had to do it the way they did it'). By contrast, children in the instrumental condition – that is, when the focus was on the material outcome – paid less attention to the sequence of pegboard actions demonstrated to them. They were also more inclined to explain what they were doing in terms of independent choice rather than social obligation (e.g., 'I can do whatever I want'). An emphasis on learning skills seemed to make the children less fastidious copiers than an emphasis on social convention.

If the desire to belong – to be one with the tribe – makes us want to copy others more closely, then we might also expect ritualistic forms of copying to intensify whenever children are worried about being excluded from the group. That is, the more strongly we yearn to belong, the more ritualistic we should become and the more faithfully we should stick to the script. To investigate these hypotheses, we carried out a series of novel experiments designed to trigger insecurity about one's position in the group, in order to explore the effects of this on ritualistic thinking and behaviour.

To do this, however, we needed to come up with an ethical and compassionate method of priming fears of ostracism in young children. Obviously, the very idea of being left out is upsetting. But the topic is often broached in children's stories, such as the song about the 'ugly duckling' who was rejected by other members of his species. In our earliest experiments exploring how fear of ostracism affected copying behaviour, we opted for an even simpler way of getting children to think about the theme of being left out. We showed our three- to six-year-old participants a pattern of shapes moving around on a video screen in a tight group formation which, when approached by a lone shape that seemed to want to join them, moved away.[20] In order to compare the effects of this with a condition in which there was no theme of ostracism, a second sample of children watched a video in which the shapes on the screen formed into groups that went merrily on their way. After watching one of these videos, all the children watched a video in which adults played with the objects: as before, they were moved around, tapped, and twirled in a video demonstration, before the children were given the objects to play with themselves. And as before, children were assigned either to a ritual condition (all objects ending up where they started) or an instrumental condition (all objects handled in exactly the same way except that one object ended up in a box, providing an obvious end goal).

The accuracy with which children copied the observed behaviours, when using the objects themselves, was then carefully measured. The results confirmed our previous findings: children copied more closely when the objects ended up where they began – making the action more likely to be seen as a ritual, with no clear causal effect on the world. But what we were most interested in here was whether fear of being ostracized might affect the way children interacted with the objects. Would children who had been primed to feel anxious about being excluded from the group copy more if the behaviour they had witnessed was more like a ritual? This is indeed what we found. Children who had been primed to think about ostracism – using our saddening video about the rejected shape – were more likely to copy the 'pointless' ritual behaviour faithfully. In other words, the fear of social exclusion made them more eager to copy in a ritualistic spirit. In addition, when asked about the exercise afterwards, children in the ritual condition who were also primed with ostracism threat were more likely to explain why they did it a certain way in terms of social prescription ('because I have to do what she does' or 'that is the way you do it') compared with those assigned to other conditions. We concluded that ritualistic copying is adopted as a way of ingratiating oneself when faced with the threat of exclusion.

While the movements of geometric shapes on a screen were capable of priming fears of ostracism, this is obviously a very mild way of doing so. In later studies, we went on to explore more visceral ostracism-priming methods. For example, we designed a study in which five- to six-year-olds were assigned to the 'Yellows' and given yellow hats and armbands.[21] After describing their favourite toys, animals, and foods, participants were told that other members of the Yellows shared those exact same preferences. Then they got to play a computer game of 'cyberball'.[22] The children were assigned a yellow avatar and either got to play the game with other yellow avatars or with avatars of a different type – the Greens. During the game, some children found that after initially being treated fairly, the avatars they were playing with suddenly stopped passing the ball, creating a visceral sense of exclusion – the ostracism prime. For others, the ball continued to be passed around fairly. Then all children in the experiment got to observe a sequence of ritualized behaviours modelled in a video, which they were told were conventional behaviours among the Yellows. We found, as predicted, that children assigned to the condition in which Yellows failed to pass the ball fairly – viscerally priming fears of ostracism – copied the ritualistic behaviours of the Yellows most faithfully afterwards. Being left out by the Greens didn't have this effect. What mattered was being left out by the group with which you identify.

All these findings pointed to the importance of ritual as a distinctively human way of affiliating with others, of establishing that we 'belong' to the tribe. Once we started to dig deeper into the question why children overimitate seemingly pointless behaviours, it became increasingly clear that they were not doing it simply to learn new skills from more knowledgeable individuals who could be trusted to not to teach them useless things. We were beginning to understand that an even stronger force was at work when children overimitate: the desire to affiliate through conformism.

There is a dark side to conformism, of course. In one of the most famous experiments in the history of psychology, Stanley Milgram showed that when subjects in an experiment are asked to administer what they believe to be increasingly painful electric shocks to a fellow subject, they will continue to do so in obedience to instructions from the experimenter.[23] Ever since, the word 'conformism' has come to be associated with unthinkingly following orders. From this, it is only a small step to think of Nazi officers carrying out atrocities on the orders of sadistic commandants. We thus associate conformism with blind obedience, loss of personal agency, and the defence that we were merely acting on orders – the proverbial 'cog in the machine'.

But the experiments my team and I have carried out suggest a more nuanced interpretation of conformism. Based on what we have learned about the urge to imitate rituals, it seems plausible that our hunter-gatherer ancestors would have been avid conformists – learning and passing on the dances, chants, sayings, and customs of their forebears. As bands of foragers moved across the landscape, joining together or splitting apart, locally distinctive cultural traditions would have formed and mutated, leaving behind a diversity of painting styles, sculptures, and other art forms. And as the first farmers constructed increasingly elaborate tombs, the differences in mortuary practices from one region to the next became more pronounced.

For example, around five to six thousand years ago to the west of where I live in Oxford, the dead were being laid to rest in stone-walled chambers under wedge-shaped mounds at lofty locations overlooking the surrounding landscape. To the southwest of me they were more likely to cover burial mounds in stone and still further west in Ireland they were building circular or oval tombs before more rectangular shapes became the vogue. These ancient peoples would have recognized each other as the bearers of distinct cultural traditions. Not only did they have locally distinctive mortuary practices but also diverse ways of decorating their bodies, behaving, talking, and marking different stages in life. And all these things were possible because of conformism bias – the willingness to copy behaviour demonstrated by trusted others for no purpose beyond the desire to affiliate.

This hints at an altogether less sinister interpretation of the human tendency to conform. While people do sometimes obey orders out of respect for authority or fear of punishment, there is also a more pervasive but sometimes downplayed cause of conformism: the simple desire to belong. We are all conformists to some extent; indeed, conformism is part of what makes us human. As I will explain later in the book, understanding the nature of conformism is vital to preventing its most destructive outcomes and to harnessing its most valuable potential for human flourishing.

2.

Wild Religion

I had been sitting all alone in the corner of the temple for nearly an hour, listening out for signs that the ancestors had come to eat our food offerings. The feast had been prepared by the village women and laid out on tables by teams of men and boys in an atmosphere of great solemnity. I had been told to listen out for signs of ancestral presence and especially for snippets of ghostly conversations that might be overheard if I concentrated intently enough. The people who built this temple were members of the Kivung – a large organization in Papua New Guinea whose members were preparing for the return of their ancestors.[1]

Through gaps in the bamboo walls of the temple, I glimpsed scantily clad bodies bustling around seemingly at random. Women hissed at misbehaving children, occasionally raising a hand that was met with cheeky grins or protestations. I wondered what it would be like to hear the voices of the dead, how you could possibly tell them apart from the low, respectful tones of the people outside. Somebody must have thrown a pile of dead foliage on a nearby fire because the smoke was now wafting into the temple, causing my eyes to smart. I also wondered what a ghostly apparition in the temple would look like. Suddenly, as I blinked away the tears, I spotted movement in my peripheral vision. Was it an ancestor? No, it was just a spider crawling slowly up a wooden post, into the rafters.

I should have known. People had already told me anyway that the ancestors were invisible. You might see signs of their presence after the fact – a morsel of the food offering that had mysteriously disappeared, for example. But you would never catch them in the act of eating. One day, however, they would suddenly appear in the village. Some said that day would arrive very soon. To be sure I understood, my informants would pinch my skin for emphasis, laughing when I flinched in surprise.

The ancestors, I was told, had originally created the surrounding rainforest. But when they returned, they would replace the trees with high-rise buildings, resembling the astonishing cities in America that everyone had seen photos of in newspaper cuttings. Only the young could read the text, and only with difficulty since those who went in search of schooling mostly gave up. Few could afford to pay with the pennies earned from cash cropping, and many told me that teaching methods in elementary schools were brutal. But the individual pages of newspapers were greatly valued for making cigarettes; they burned more readily than banana leaves as a skin for home-grown tobacco. Before rolling up the paper and burning it, however, smokers pored over the pictures in newspaper articles, especially those depicting city life. Their vision of paradise was one in which such tower blocks would emerge magically overnight, affording forms of unimaginable opulence and comfort assumed to go along with urban lifestyles. There would then follow a day of judgement in which the wicked would be cast into hell and the faithful would embark upon a new life of ease and freedom from suffering for the rest of eternity.

In the meantime, I sat patiently at my post, listening for the sounds of ghostly presence. I soon found myself contemplating another puzzle. Why did people everywhere in the world entertain the idea that we live on after we die, that we are surrounded by spirits that float around invisibly, that supernatural beings created the world around us, or that they can be gratified by prayers or offerings? Even then, at an early stage in my efforts to learn about the nature of religion, it was beginning to dawn on me that there was a common bedrock of beliefs and practices shared by humanity at large. But why?

Decades later, I have come to appreciate that the fundamental building blocks of religious thinking and behaviour are found in all human societies around the world – from small-scale indigenous communities to industrial conurbations and inner cities. For example, people everywhere imagine that bodies and minds can be separated, that features of the natural world have hidden essences and purposes and that we ought to please and placate the spirits of the dead. People everywhere are inclined to spread stories about miraculous events and special beings that contradict our intuitive expectations about the way the world works, giving rise to a plethora of beliefs in supernatural beings and forces of various kinds. People everywhere – even babies who have not yet learned to speak – experience feelings of awe and reverence when they encounter those who can channel otherworldly powers.

The recurrence of such beliefs points to a pervasive and panhuman 'religiosity bias'. As we will see later in the book, this bias not only shapes the things we think of as 'faiths', 'superstitions', and 'fairy tales' but also various contemporary forms of advertising and consumer behaviour. But for now, let's focus on why our religiosity bias can be seen as a fundamental aspect of human nature.

Missionaries, spirits, and ancestors

I first began thinking about our bias to believe during my time as a young doctoral researcher living among the Mali Baining[2] – one of many language groups that had joined the Kivung organization as it spread through the rainforest from the mid-1960s and which served to unite them into a new political force in the region.[3] It may be tempting to assume that all the ideas I encountered in the Kivung had their origins in European influence – and especially the teachings of Catholic priests and catechists. These missionaries began their patrols among the Baining people long before my own arrival. There, in the scattered villages of the East New Britain rainforest, they built churches, told stories from the Bible, conducted baptisms, and heard confessions. And it is certainly true that many Kivung ideas were influenced by priests from overseas – especially those concerning heaven and hell, sin, and salvation. But part of my job as an anthropologist living with the Baining was to document the details of a much more ancient cultural system, which preceded the arrival of Europeans and their strange beliefs. I therefore spent many of my days in lengthy conversations with elders who remembered a time before the teachings of missionaries were widely known and when indigenous religious beliefs and practices were still revered and closely observed.

In ancestral times, adulthood was achieved not by simply letting nature take its course but by requiring boys and girls separately to undergo a very elaborate and challenging process of initiation. This involved long periods of seclusion at sacred places in the forest. These periods were punctuated by traumatic ordeals so secret that divulging them could result in being put to death.[4] It was through these rituals that the entire cosmos was thought to be regenerated. They were believed to be necessary not only to enable children to grow and mature but also to ensure an abundance of creatures and plants in the forest that were vital to human survival, reproduction, and flourishing.

To help me understand this elaborate system of beliefs, I was eventually taken deep into the rainforest to discover for myself the secrets of the male cult. As part of my initiation into manhood, the elders taught me about everyday aspects of the supernatural world around us. They explained that the forest was teeming with invisible spirits, who could be persuaded to assist with various human endeavours, from hunting to healing.[5] But the same spirits who could be pushed and cajoled into helping us were also capricious and easily offended. They needed to be continually placated and indulged through offerings and sacrifices. On one occasion, as I thrashed my way clumsily through the jungle with my bush knife, I was told I had disturbed the garden of a tree-dwelling spirit, mistaking it for wild undergrowth. As a result, I was encouraged to perform various magical spells, to ward off punishment and restore cordial relations with the spirit world.

People referred to such beings as *sega*, meaning 'spirits of the forest'. Although invisible to the human eye, *sega* were often seen in dreams – particularly the dreams of spirit mediums known as *agungaraga*. These were special individuals who not only communed with the *sega* in their sleep but who, as a result, frequently learned the hidden causes of people's misfortunes, such as unexpected crop failure or a sudden illness. The knowledge of *agungaraga* was held in high esteem: they provided a source of practical solutions to everyday problems, particularly by performing magical rituals. Very rarely, though, people encountered *sega* in waking life – an experience always described as eerie and in some cases terrifying. Such encounters were invariably accompanied by some inexplicable feature that defied familiar laws of physics. For example, the approach of the *sega* might be perfectly silent and thus undetectable to the human ear. Or the apparition might disappear the very instant one averted one's eyes.

These ideas might sound alien to western ears. But in fact, traditional Baining beliefs are strikingly similar to those found in all the world's cultural systems – from children's stories and songs through to the most sacred teachings of the world religions. Whether you are an adherent to some form of Christianity, Islam, Hinduism, Buddhism, Taoism, Zoroastrianism, or any other of the major forms of organized religion, you will most likely subscribe to some kind of belief in the supernatural, in life after death, in the hidden powers of ritual, in intelligent design of nature or the cosmos. These ideas are rooted in deep intuitions, finding expression not only in doctrinal belief systems curated by monks, priests, gurus, and other religious specialists but also in more informal varieties of religious life – from spiritualism and New Age beliefs to shamanism and possession cults.[6]

Given these similarities, it may come as a surprise to discover that the Catholic missionaries patrolling the Baining rainforest took a dim view of the *sega*. It was not simply that the *sega* did not fit into the cosmological framework of the church, and could thus be dismissed as childish fantasy equivalent, say, to beliefs in fairies or Santa Claus. On the contrary, successive priests and catechists visiting the village declared the *sega* to be perfectly real but evil – emissaries of Satan. Why?

The answer to this question is undoubtedly complex and multifaceted but it breaks down into two main aspects. First, the Roman Catholic religion is a system of beliefs and practices that adherents are supposed to adopt as a complete and somewhat exclusive package. Ideally, you should not be 'a bit' Catholic – for example, only on Sundays – and then an animist for the rest of each week. Priests therefore commonly integrated local beliefs into the Vatican package but in a way that disincentivized engagement with them. Second, the approved supernatural agents of the missionaries – the Holy Trinity of Father, Son, and Holy Spirit – were morally concerned beings. The God of the missions cared about human affairs in ways that the *sega* did not. Paradoxically, the best way to exclude the *sega* from the Catholic belief package was to suck them into its moralizing framework; the *sega* could in this way be roundly condemned rather than venerated, petitioned, or worshipped alongside the sacraments bequeathed by Rome.

By contrast, from the time of its foundation in 1964, the indigenously led Kivung movement actively encouraged followers to take pride in their traditional beliefs and practices and to maintain harmonious relationships with the *sega* rather than to spurn them as satanic. Consequently, whenever I inadvertently offended the *sega*, people encouraged me to repair the damage. Usually, the cause of the offence was that I had encroached on their territory or damaged their property. I never incurred the wrath of the *sega* by behaving badly towards other people in my community. I was assured that the *sega* could not care less if I wronged my neighbour by stealing from his gardens or running off with his wife or even murdering his brother (not that I did any of those awful things, obviously!). What really upset the *sega* was if I interfered with their lives directly. As long as I avoided doing so, I could do whatever I liked for all they cared. Likewise, you couldn't persuade the *sega* to do something for you, such as help you hunt successfully or recover from an illness, simply by acting morally in your own community.

Instead, this could only be achieved by doing something of direct benefit to the *sega* – such as offering them a sacrificial gift or bribe – or by uttering spells that compelled them to obey your command. Even then you could never be sure it would work; you might get the spell wrong or simply get the *sega* on a bad day. But what you could be sure of is that the spirits of the forest were only interested in themselves and what you could do for them, not whether you were a good person.

This is where the Abrahamic God of the missions genuinely did introduce a radically new idea to the Baining rainforest. The missionary teaching that God cares what you do to others in your community – punishing sins and rewarding piety – came as a revelation. Even more astonishing was the idea that the God sees into your soul and will punish your bad intentions, even before they are enacted. These ideas became fundamental to the Kivung but only after extensive discussion. In later chapters we'll see that beliefs in moralizing supernatural punishment emerged quite late in the evolution of complex societies. This is because supernatural agents are not intuitively regarded as enforcers of morality in human affairs. We more naturally think of them as powerful beings we had better keep on our side, like the *sega*. Consequently, in small-scale societies, spirits and gods are basically a pantheon of egotists – how we behave among ourselves is up to us as long as it doesn't get in the way of our ability to placate, bribe, and flatter them.

So, the answer to the question of whether the Kivung was an indigenous belief system or one borrowed from the Christian missionaries is that it was both. The Kivung adopted many of the ideas of the Roman Catholic mission pertaining to sin and salvation, attributing such moral concerns to local ancestors. But it also encouraged people to maintain their traditional beliefs in the capricious *sega*, with whom cordial relations needed to be preserved. This kind of 'mix and match' approach to indigenous and Christian beliefs is common in colonized regions where missionaries set up shop.

As a young doctoral student, I found the mixing and matching of these beliefs fascinating. It seemed odd to me that ideas from a wholly alien theology – Catholicism – could be so readily integrated with indigenous beliefs to create the Kivung worldview. And this got me thinking about whether facets of both clusters of religious beliefs were in some way *intuitive*. What if there was something inherently appealing about both Christian and indigenous ideas – that they contained aspects of religious thinking and behaviour that are shared by people everywhere, and which arise and spread spontaneously in almost any group?

Religion in the wild

Atheists like to point out the absurdity of declaring one unique system of beliefs to embody ultimate truth while all the others are false. After all, the usual reason for such special pleading is that you just happen to have been raised in that tradition (rather than that you systematically compared every religion going and found only one of them to be correct). On those grounds, religious convictions might not seem quite so absurd if we could somehow strip away all the elements that make one religion different from another. If we could discover a common core to all religions, then cleaving to those core beliefs would simply be an act of agreeing with everyone else who subscribes to some sort of religion. Although this wouldn't necessarily make such beliefs more valid, it would lessen the absurdity of adopting one variant in preference to another purely on the grounds of one's upbringing.

My time with the Kivung set me on a path to search for such common features. These features have together been described as 'wild religion'[7] because they comprise a set of beliefs that are rooted in universal predispositions to think in certain ways. Such beliefs are different from the more elaborate and distinctive doctrines of organized religions. The latter are often referred to as 'doctrinal' aspects of religion, requiring explicit teaching and training, consultation of sacred texts, or frequent repetition of dogmas and narratives.[8]

Doctrinal religion is a manicured garden, which needs to be curated and tended by experts. But wild religion spreads everywhere quite naturally, like weeds that germinate without assistance. This might explain why the universal elements that make up wild religion are so often regarded as a nuisance, at least by religious authorities.

Still, it is hard to imagine any religions – doctrinal or otherwise – without these wild aspects built deep into their makeup. For example, although an organized religion like the Roman Catholic Church is continually having to fend off 'wild' ideas about witchcraft, spirit possession, ghosts, and black magic, it also tolerates and even encourages authorized variants of those kinds of beliefs as part of everyday practice: hence the existence of weeping statues and healing relics. In other words, although the central doctrines may be quite unnatural and hard to learn – from the mystery of the Holy Trinity to the Noble Truths of Buddhism – lay adherents of organized religions are given plenty of scope to indulge in 'folk' traditions and cults that incorporate much 'wilder' and more naturally intuitive elements.

Claiming that wild religion comes naturally is not the same as claiming that we are born with a fully formed set of religious beliefs – much like saying that underarm hair is natural does not amount to saying we are all born with hairy armpits. Rather, the idea is that infants have deeply ingrained propensities to acquire wild religious beliefs over the course of their childhood, as long as the environment around them provides enough relevant material to work with and suitable conditions for sponging it up.

European notions of witches are rooted in evolved intuitions too. But with the right cultural conditioning, these intuitions can be overridden. If you have been raised in an atheistic community – one that systematically discourages the default propensity to believe in supernatural beings – then you will likely be sceptical that witches exist. For this reason, it is important to emphasize that describing wild religion as 'natural' – one of our evolved human biases – does not mean we are born with it, nor that it will inevitably find expression in our cultural beliefs. These intuitive beliefs are, however, unusual in that they do not need to be explicitly taught; they arise naturally from the way our brains typically develop in standard environments.

To understand how, it is worth reflecting briefly on how the human brain develops. The brain is not simply a blank slate on which life experiences are mysteriously inscribed. One view is that it is a highly specialized computational device – or rather a vast array of such devices. That is, just as bats have dedicated machinery for mapping their location in the physical environment using sonar, humans have dedicated cognitive architecture for interpreting other people's mental states, for anticipating the trajectories of objects moving through space, and for reasoning about the essential differences between species. In other words, we don't have to be explicitly taught that people can be deceived or that animals come in different varieties – we believe those things in part because of the way our brains evolved to process information about the kind of world we and our ancestors have always lived in. On this view, the human brain is a bit like a Swiss Army knife, with lots of different blades designed for different applications (not just cutting through softer things but pulling corks out of wine bottles or prising stones out of horses' hooves).

Another view is that the brain is more like a general-purpose tool – less like a Swiss Army knife made up of different blades with different functions, and more like a single blade that can be used in a great variety of ways as the need arises. According to this view, the many specialized tasks that the brain can accomplish are largely a product of distinctive learned techniques for handling a blade. To take an obvious example, the brain does not have evolved circuitry dedicated for reading and writing. Literacy is more like a cognitive gadget that has to be learned by each successive generation of readers and writers. Of course, literacy appeared too recently to have evolved under natural selection and it doesn't arise at all unless you grow up in a group in which it is practised. The question is the extent to which other such gadgets, which are more plausible candidates for specialized evolved capacities – such as a set of universal grammatical principles – might actually be a consequence of learning rather than being innate.[9]

Both perspectives on how the brain works are quite plausible and potentially reconcilable. In both cases, the brain has an in-built potential for certain kinds of tasks but can also be flexibly applied to a host of new situations – ones that were unimaginable in our evolutionary past. Whether we emphasize the role of innate specialized psychological tools or learned cognitive gadgets, our evolved psychology surely combines both of these capacities – and does so in ways that influence the kinds of cultural beliefs we are likely to sponge up and pass on.

To the extent that certain features of religiosity bias are universal today, it is likely that many of those features were also universal in the human past. Although it is impossible to reconstruct the contents of belief systems in ancient prehistory in any detail, archaeology provides direct evidence for at least some features of panhuman religiosity. For example, the presence of grave goods in so many prehistoric burials strongly suggests that modern humans (perhaps among other closely related species, such as Neanderthals) have always entertained hopes for an afterlife, and paintings on cave walls arguably point to beliefs in the presence of supernatural beings of various kinds. Moreover, the fact that so many religious intuitions emerge early in childhood suggests not only that humans have always believed such things but will probably also be inclined to do so in the future. At least some aspects of religion, then, come naturally. Psychologists describe these natural aspects as 'intuitive'.

The intuitive foundations of wild religion

When I returned to King's College, Cambridge after living for two years in Papua New Guinea, immersed in the world of the Kivung and the *sega*, I found it hard to reintegrate. The norms and conventions of my tribe now seemed oddly alien to me, having been thrown into sharp relief as a consequence of living in another.

When I first met Pascal Boyer, however, I sensed that he was like me in this respect. He was a research fellow in my college. He lived in an enviable apartment spanning the arch of Gibbs Building overlooking the 'the backs' and the River Cam in the distance. By contrast, I was still just a research student, writing up my doctoral thesis, shuffling around the library in crumpled jeans, reeking of roll-ups. But I think we saw each other as kindred spirits, having both returned from fieldwork in faraway cultures and having been struck just as much by the similarities of humans everywhere as by their differences.

As we sat begowned and besuited at college dinners, we exchanged ironic grimaces across the table about the insistence on passing port in one direction rather than another and shared anecdotes about the similarity of things we had observed beforehand in the college chapel and magical practices we had studied in distant lands. Boyer had conducted fieldwork in sub-Saharan Africa, among the Fang of Cameroon, on the other side of the world from the people I lived with in New Guinea. But we agreed that our experiences of life in decidedly non-English villages, surrounded by beliefs in vengeful witches and rapacious ancestors, were nevertheless rooted in much the same psychology as the squabbling college fellowship and deference to tradition at Cambridge high table.

Boyer soon introduced me to some of his friends in America and before long we were all meeting regularly on both sides of the Atlantic, joined by a growing cabal of bright young scholars.[10] Although the ideas of this group were diverse – often resulting in discussions that went on late into the night – they were always good-natured, humorous, and perhaps even more importantly, premised on at least two explicitly agreed principles.

The first principle was that religion could be explained scientifically. The second was that the explanatory power of scientific theories mattered more than who came up with them. Both principles were quite radical in the disciplines in which most of us had been working. At that time, many of our colleagues in the departments of religion, anthropology, and other humanities disciplines were sceptical or even openly hostile towards scientific methods which they alluded to, shudderingly, as 'reductionism'. And exponents of the dominant frameworks in those disciplines were often proprietorial and defensive about their theories. Being part of a group with a more scientifically grounded and collaborative approach was more than just refreshing – it was life-changing. Our group eventually came up with a catchy name for this new branch of academic research. 'Cognitive Science of Religion' – or CSR as it came to be known – was an attempt to explain recurrence and variation in religious belief and behaviour around the world based on a rapidly growing understanding of our cognitive architecture.

Explaining those 'wild' patterns of religious thinking and behaviour that pop up and spread naturally in all human societies rapidly became the main goal of the new CSR field.[11] And for the last several decades, its practitioners have been gradually unpicking the psychological processes that give rise to religion in the wild. This research suggests that many widespread features of religious thinking and behaviour are best understood as *by-products* of evolution. While most wild religious beliefs were not selected for in the human evolutionary journey, they came about as a side effect of other, more adaptive psychological characteristics: for example, our intuitive psychology (including the way we anticipate how others will behave), our intuitive biology (such as the way we create taxonomies of the natural world), our tool-making brains (which make us think everything has a purpose) and our tendency to overdetect agency in the environment (particularly our early warning systems for detecting predators). These intuitions helped our ancestors to survive and pass on their genes, but they also make us naturally susceptible to believing in an afterlife; they make us treat certain objects or places as sacred; they lead us to think that the natural world was intelligently designed; and they convince us that invisible and dangerous spirits lurk in caves and forests.

For example, it seems that several recurrent features of religion in the wild are a by-product of the human capacity for reasoning about the mental states of others. Experimental psychologists refer to this as 'mindreading'.[12] They don't mean that we engage in telepathy – just that we naturally try to imagine things from the perspective of other people and to predict their behaviour based on what we think they know or don't know. Even a fairly crude version of this capacity is unavailable to toddlers, who simply assume that if they have been shown where an object has been hidden, everyone else knows what they know. By age five, however, children will typically realize that if you have not been shown or told where an object is hidden, you will not know it exists or where to look for it. Some five-year-olds are amusingly devious and – for better or worse – derive a great deal of satisfaction from testing their new-found ability to tell fibs. They have acquired what psychologists call 'theory of mind'.[13]

This ability to reason about the mental states of others has profound consequences for our ideas about spirits and the afterlife. A good example of this is the so-called 'simulation constraint hypothesis' – an attempt to explain beliefs in the afterlife as a side effect of the way we naturally reason about minds. According to this hypothesis, the idea of a ghost or spirit of the dead results from the impossibility of imagining the elimination of certain mental states. Of course, we can imagine what it would be like to lack certain perceptual capacities, such as sight and hearing, simply by covering the relevant organs (eyes and ears) with our hands. But it is far less easy to simulate the lack of higher-order cognitive capacities, such the ability to remember, emote, or reason. The moment you try to imagine those capacities being switched off you realize how hard that is. Far more natural, so the argument goes, to imagine those things persisting after death.[14]

In other words, we imagine the spirits of the dead to remember things that happened during their lives, to have feelings, and to form judgements even if we know that their bodies (including their eyes and their ears) no longer function and may be incinerated or rotting in the ground. This theory is consistent with the very early development of such assumptions in children. For example, even preschool children believe that the higher-order cognitive functions of a puppet mouse persist after he has been gobbled up by a puppet alligator.[15] The little mouse may no longer need to go to the bathroom or to eat breakfast, but children readily assume he can still remember his mother and his friends even though his mousey body is no more.

Another intuitive foundation of wild religion stems from the way we reason about catching diseases. Long before scientists knew about microbes, people worried about the risks of contagion. This was not only because they observed the way symptoms spread through contact. It was also because they intuitively reasoned that invisible essences are present in the world. There are obvious evolutionary benefits to this intuition: if you naturally fear contagion through contact with contaminants, you may be less likely to catch a virulent and deadly disease spreading through your community. This might explain why certain hazard precaution routines are standardly observed in all human populations, such as cleaning, straightening, and separating certain objects. Interestingly, these are also the same precautionary routines that become tragically exaggerated among sufferers of obsessive-compulsive disorder.[16] But although such concerns can become pathological, we all have a certain degree of OCD-like concerns about hygiene and contagion.[17]

But this intuition too has religious side effects. Fears of contamination by hidden essences are commonly expressed not only in rituals but also in fears of sorcery and black magic. Across many different cultural groups in Papua New Guinea, for example, I encountered the idea that one should be careful not to leave behind traces of self, such as items of clothing, hair, or nail clippings, when interacting with people you do not trust. Such items contain your essence and could therefore be used by your enemies to cast evil spells on you, even though the rest of your body may be far away. In many cultures, the idea that our hidden essence leaves traces on the things we touch is given a name. In many Oceanic groups, for example, the word for it is *mana*, whereas in Madagascar it is *hasina*, in Objibway it is *manitou*, in Iroquois it is *orenda*, and so on.[18] As a result, the idea of magical contagion has fascinated scholars for generations, giving rise to a profusion of different theories about it.[19]

Linked to this is the idea that people's positive and negative qualities are inherently contagious. For example, if a person is revered or worshipped, then touching or kissing an object they have worn or that has been fashioned in their likeness may bring us benefits, such as health or good fortune. But the converse is an even more common idea: people of low social standing, especially those excluded or ostracized for moral failings or transgressions, should be avoided like the plague. Literally. In Papua New Guinea, the bodies of hated enemies were often thrown into rivers to wash them away. In most contemporary states, convicted criminals are locked away with the result that law-abiding citizens don't have to see them or consort with them. The more extreme a person's crimes, the more completely we may seek to obliterate all traces of them. In the UK, houses of serial killers depreciate sharply in value and are often demolished. In extreme cases, such as the house occupied by Fred and Rosemary West who raped, tortured, and murdered young girls, the rubble from their house was crushed into dust, and then disposed of at a landfill site away from human habitation. People everywhere seem to feel a spontaneous aversion to having physical contact with objects associated with evildoers – even to the extent of being reluctant to don a murderer's sweater in psychological experiments.[20] Fears of contamination by the forces of darkness tend to be more widespread in populations where communicable diseases are prevalent, further supporting the hypothesis that this aspect of our bias to believe in evil essences is a by-product of our evolved psychological defences against contagion.[21]

Yet another intuitive foundation for wild religion is a panhuman bias known as 'promiscuous teleology'.[22] Teleological reasoning is the tendency to see the hand of a creator everywhere – to conclude that everything from trees to rocks have been actively designed that way. Humans tend to apply teleological reasoning willy-nilly to just about everything, including the natural world. This promiscuous application of teleological reasoning is another feature of human psychology that emerges quite early in development. For example, when asked whether rocks are pointy due to long-term erosion or to stop elephants sitting on them, preschool children prefer the teleological explanation – one that attributes a functional design to the shape of the rocks.[23] The natural human tendency to reason teleologically is of course highly adaptive, because it enabled humans to creatively develop tools that allowed them to kill animals at a distance, to cut through their hides, to start fires to cook the meat, and to do countless other things that would have been impossible if relying on our puny bodies alone. But a by-product of this teleological mindset was the tendency to see creator beings everywhere. As such, this bias may help explain the pervasive recurrence of creation myths in the world's religions, from the Dreamtime heroes of Indigenous Australians to the genesis stories of Abrahamic religions. In trying to explain the world around us, our natural tendency is always to imagine that some divinity or ancestor made it that way and to make up and spread stories about how it happened.

But for me, the most intriguing intuitive belief of all relates not to the afterlife, creationism, or contagion – but to our tendency to see agents lurking all around us, even when we're all alone. This tendency grows stronger whenever we are in a potentially dangerous or unfamiliar environment. Given that we are naturally inclined to entertain ideas about spirits that live on after we die and powerful invisible beings that live in the trees, mountains, and rivers, it is hardly surprising that certain places make us a little edgy. Consider how it feels to be lost in a darkened forest or a graveyard as the sun goes down. A sudden fall of leaves or the hooting of an owl is not only likely to make us jump but also to plant the idea in our minds that there are humanlike agents around. To the extent that our culture furnishes us with a pantheon of malevolent as well as benevolent beings, such experiences can be quite frightening. When things go bump in the night, our thoughts naturally turn to ghouls, vampires, witches, and devils.

Psychologist Justin Barrett has argued that such apparitions are generated by a Hypersensitive Agency Detection Device, or

'HADD', buried deep in our brains.[24] Having a hair-triggered device of this kind would have been useful to our ancestors when surrounded by powerful nocturnal predators. Better to overdetect the presence of a hungry lion approaching the camp under cover of darkness and needlessly run for the trees after a false alarm than be too nonchalant and get eaten alive. The HADD idea explains not only our fear of the dark but specifically the tendency to interpret mysterious events as signs of agency. Given that the most threatening predators in human prehistory were not lions but other humans, this may also reinforce our universal tendency to see human forms in otherwise random shapes. Hearing a twig snap in the darkened forest, for example, is arguably not just about the overdetection of agency but more specifically the overdetection of *humanlike* agency – a trait that is usually described as 'anthropomorphism'. This might go some way to explaining why humans everywhere are so prone to spotting supernatural forces lurking in trees, rivers, and rocks; there is a strong evolutionary benefit to attributing humanlike qualities to our surroundings.[25]

All in all, then, certain ideas about the supernatural would seem to arise naturally because they are rooted in our evolved intuitions, for example about how the world around us came into being and what to do when we hear an eerie sound in the dead of night. But this is not the only form 'wild' religious beliefs take. There is evidence that some beliefs about the supernatural spread not because they are intuitive, but because they are the opposite. This is the perplexing domain of minimally counter-intuitive beliefs.

Catchy counter-intuitions and their social consequences

My parents' flat was bustling with guests. One of my father's colleagues had come to visit for Sunday lunch, together with his new wife, who had recently arrived from Hyderabad. The colleague's son Sanjay was also in tow, and I had been allowed to invite my two best friends from primary school in the hope that we would entertain him. I had a lurking suspicion that Sanjay considered my family beneath him on the social ladder because we were much poorer. On the other hand, I had heard that his father was anxious we would judge him, by the lights of English standards, for agreeing to an arranged marriage after his last wife died. I wondered if Sanjay was still in mourning. The atmosphere lightened when my father's colleague clapped his hands together and told us about a technique of levitation he had learned from his uncle in India.

The technique was simple. One of us sat on a chair while the rest of us gathered around and placed our hands above the head of the person seated. Our hands should not touch but be held just an inch or two apart, one above the other in a coordinated fashion vaguely reminiscent of a seance. And then, with great solemnity, one index finger from each of us should be placed under the crook of the seated person's knees and armpits. Slowly those fingers should then rise up, hoisting the seated person aloft so that they floated effortlessly above the chair.

Looking back now, my schoolfriends and I experienced this demonstration with a mixture of awe and scepticism. On the one hand, we agreed that the body of the sitter did seem to be extraordinarily light. It was certainly not weightless. But, on the other hand, it was not as heavy as any of us expected. None of us could have imagined that our fingers would be strong enough to lift an entire person. We tried to compare the effects of lifting someone without first performing the ritualistic preamble. One of my schoolfriends insisted that the levitation effect only seemed to work if we followed the stipulated magical sequence correctly. I couldn't help wondering, though, if he was putting in the same effort to lift when we tried to perform the levitation without the magical preamble. Even to my childlike mind, this did not seem to be a well-controlled experiment. I went along with it because what mattered socially was the shared excitement, the feeling that we had become channels for something powerful but inexplicable.

Ideas about levitating or floating bodies crop up in lots of different cultures. But why? A potentially compelling answer is that such ideas directly contradict our natural intuitions about the way the world works – in this case our early developing assumption that unsupported objects fall earthwards – and that makes such ideas 'catchy'.[26] Humans' assumptions about gravity are detectable at a remarkably young age. Developmental psychologists have shown that early in infancy, babies expect objects to fall to the ground if they are dropped and express surprise if an object floats upwards or hangs suspended in space without any supports.[27] These assumptions about how gravity should work seem to be built into babies' cognitive architecture. A psychologist would say these assumptions are 'underspecified by experience' – meaning that babies haven't had enough exposure to the world to have predicted the effects of gravity based on observation alone. They seem to be naturally predisposed to anticipate that objects will behave in that way and to find floaty things, such as balloons filled with helium, counter-intuitive.

But how, you may ask, could we possibly know what babies expect objects to do? Obviously, you can't just ask them because they are too young to speak. But what developmental psychologists *can* do is establish which scenarios, presented in a carefully controlled lab setting, babies regard as more or less surprising. There are two ways in which surprise is typically measured in these situations. One is looking time: infants stare longer at an event they weren't expecting, indicating curiosity. Another is sucking: babies suck harder on an object in their mouths when they are surprised and therefore engrossed by an observed event. In numerous experiments, infants have been shown to exhibit expressions of surprise when objects behave in ways that are physically counter-intuitive (such as floating upwards, popping up out of nowhere or passing through each other like a ghost).[28]

This early developing fascination with counter-intuitive events is no doubt why children's stories are full of them. Even young children are naturally absorbed by stories in which Superman can fly and Santa can pass through unfeasibly small chimneys. But the allure of counter-intuitive beliefs continues well into adulthood.

This might explain not only why fairy tales spread so widely and are remembered and cherished across the generations, but also why ideas about supernatural agents disperse and endure in folklore.[29] In adulthood, these beliefs tend to remain counter-intuitive without actually being *surprising*. Well after we have become familiar with the idea of flying objects – from aeroplanes and gliders to birds and balloons – defiance of gravity is no longer unexpected but is still eye-catchingly counter-intuitive. Our tacit intuitions still inform us that the jet plane should really fall out of the sky, which is why many of us still feel that flying is vaguely miraculous, and some would rather not risk it at all.

The appeal of counter-intuitive beliefs goes some way to explain the prevalence of supernatural beings in religious texts and myth cycles. Many of the world's supernatural beings – such as ghosts, spirits, gods, and ancestors – are counter-intuitive in much the same way as aeroplanes and hot air balloons. In fact, if you want to create a simple ghost concept, just try to think of somebody who died in your house many years ago but occasionally returns in the dead of night, popping up out of nowhere in the attic or floating in and out of the windows at will. The apparition you have now conjured up is just like any other human being except that they can teleport and defy gravity (and come back from the dead – but we'll get to that later). It only requires one or a few counter-intuitive properties to turn a perfectly normal conception of a person into a ghost, spirit, ancestor, or deity.

It has been argued that most of the supernatural beings found in the world's many cultures can be predicted on the basis of a remarkably limited number of counter-intuitive tweaks to otherwise perfectly natural agents.[30] According to Pascal Boyer, two kinds of 'tweaks' commonly underlie concepts of the supernatural, building on our evolved psychology.[31] The first is simply to take some intuitive property associated with a particular domain and specify that this property does not apply. In the example I mentioned in the last paragraph, a simple ghost concept could entail violating expectations pertaining to all physical objects, by describing scenarios in which a person can teleport or fly or pass through walls. Such scenarios defy intuitive physics – in particular, our naturally developing expectations that all objects must move through space in continuous paths (not pop up out of nowhere), fall earthwards if unsupported (not float upwards or fly), and displace or break other objects if colliding with them (not pass through solid objects). Boyer refers to such counter-intuitive scenarios as 'breaches' of intuitive knowledge. The second kind of 'counter-intuitive tweak' would be one that takes a property from one domain and applies it to another. For example, a weeping or bleeding statue transfers expectations from the domain of intuitive biology to the domain of artefacts.

What's crucial, though, is that these supernatural phenomena are only *mildly* counter-intuitive. We may believe in a weeping statue; we tend not to believe in a statue that only exists on Wednesdays. Boyer refers to the milder versions as 'minimally counter-intuitive' – a phrase that is now widely referred to in the literature as 'MCI'. A veritable industry of research on the causes and consequences of MCI concepts now exists, much of it focused on the question of whether MCI concepts are more widespread cross-culturally because they are 'catchier' than merely intuitive ideas.[32]

This catchiness would explain why such kinds of religious constructs are continually reinvented across remote cultural systems – from the primordial beings of ancient myths to the superheroes of Marvel comics, and from the accounts of miracles in the New Testament to children's fairy tales. A central idea here is that, although the supernatural beings we treat with solemnity and respect in places of worship may prompt very different reactions from those depicted in children's stories, they are rooted in the same basic psychology. Declaring that Santa knows who has been naughty or nice may prompt a smile while denying the infallibility of the Pope could – in certain historical periods – get you burned at the stake. But the psychology needed to entertain both ideas is not so different.[33] This would help to explain how both ideas initially got off the ground historically, irrespective of their contrasting social consequences.

By the early 2000s, research on MCI concepts was rapidly

expanding. More and more questions were being raised not only about the cognitive catchiness of supernatural beliefs but their social consequences. Do such beliefs help or hinder cooperation? Do they support or undermine the social order? And above all, are we naturally inclined to bow down to the gods and their earthly emissaries – and how might we tell?

The gods are in charge

One of the most widespread social consequences of religion in general is that – for better or worse – it tells us what to do. Around the world, ancestors, gods, and supernatural beings of all kinds are not merely our equals but are generally seen as beings we should defer to. If we offend them, we must make amends. If we want their forgiveness, we must grovel before them. If we want their help, we must flatter them or shower them with gifts. Whether in the role of bully or beneficent provider, the gods are always in a position of dominance rather than servitude. The gods boss us around. Our role is to plead for mercy or try to please them.

My colleagues and I have been trying to explain why this is. Do we *naturally* expect MCI beings (e.g., gods, spirits, ancestors) or their earthly emissaries (e.g., prophets, priest, gurus, and messiahs) to command deference – requiring us to bow down to them both literally and metaphorically?

The answer, it seems, is yes. We have evidence that even very young babies expect supernatural beings to be socially dominant. Explaining how we have come to this conclusion, however, requires a bit of unpicking. It begins with an observation about babies' attitudes to authority. As well as their fondness for counter-intuitive concepts, preverbal infants show a remarkable sensitivity to information about who is the boss. For example, when watching animated cartoons, they expect characters who have relatively larger body-size, more allies, or who are located at a higher elevation (e.g., standing on a hill) to prevail in situations of conflict. They also express surprise when an individual they assumed to be the boss fails to get what they want in a standoff. These expectations appear to arise naturally in the early developing human brain and they also influence the way our minds mature, adapting to whatever cultural environments we happen to be raised in. For example, when we meet television personalities in the flesh, we are often surprised that they are not taller, presumably because we expect high status (associated with fame and fortune) to be correlated with physical stature and formidability.

A key question is whether babies might attribute authority to supernatural beings in the same way. After much discussion of this question, I teamed up with psychologists at Kyushu University in Japan to draw upon their specialized facilities for conducting studies with preverbal infants.[34] We set out to design an experiment aimed at establishing whether these early developing intuitions about social dominance might be directly related to the way we think about religious authority. Could it be, for example, that infants expect individuals exhibiting supernatural abilities – such as the ability to defy gravity or pop up out of nowhere – to be socially dominant, in much the same way as a physically larger or more muscular character? If yes, could this explain why throughout human societies, leaders who are thought to wield supernatural powers – from shamans and witchdoctors to priests and popes – tend to be accorded higher social status?

To investigate these questions, our research team invited parents to bring their babies into the lab to view a series of animated videos in which two characters competed for a reward. In each scenario, one character progressed across the screen in a counter-intuitive way – either flying or teleporting in the direction of the reward. In contrast, the other character in the video moved more intuitively across the screen by sticking to the ground and travelling in a continuous path from left to right. The first character therefore had supernatural powers; the second was sorely lacking them. Next, the two characters found themselves in a standoff. Placed between the supernatural agent (the one who could levitate or teleport) and the natural agent (the one who could do neither) was a green cube. Not exactly the most desirable reward imaginable, you might think. But from the way in which each of the agents on the screen moved towards the green object in incremental steps, uttering grunts that indicated effort and desire, it was clear each of them wanted to take possession of it. There could only be one winner, however. And we wanted to find out which outcome the babies would find most surprising. Sure enough, after the final standoff was resolved, the babies stared longer – indicating surprise – when the *natural* agent won the reward. Babies were apparently expecting the supernatural agent to exert social dominance by taking possession of the little green cube. They were confounded when that didn't happen.

This was the first study to show that even preverbal infants expect beings with supernatural powers to be socially dominant. It was an exciting finding because it suggested that even before we have learned much about the religious beliefs of our group, we expect the gods (or their earthly intercessors) to be in charge, and not the other way around.

Religion and morality

There is another crucial aspect of human nature that resurfaces in strikingly similar ways cross-culturally: our moral compass. Although it has nothing much to do with weeping statues or cube-snatching superheroes, many people nowadays connect such ideas to beliefs in higher powers. If asked what the most important social consequences of religion are, many people would say it is that religious beliefs make us act more morally. A survey conducted by the Pew Research Center in 2007 showed that in answer to the question 'Do you need to believe in God to be moral?', the overwhelming majority of people in countries outside Europe said yes.[35]

But surprisingly, perhaps, scientists remain divided on the question.[36] Part of the reason for this is that there are many gods and many moral systems, and it isn't always clear what we mean when we refer to either religion or morality. Nevertheless, one might still ask whether morality is in some way integral to religiosity bias. Do our intuitive ideas about the afterlife, contagion, intelligent design, or even MCI properties fundamentally alter our moral behaviour? Socrates posed a similar question when he asked whether goodness is loved by the gods because it is good or whether goodness is good because it is loved by the gods.[37]

Today, some of the best answers to this question come not from Greek philosophy but from scientific research. Studies led by my colleague Oliver Scott Curry have shown that much of human morality is rooted in a single preoccupation: cooperation. More specifically, seven principles of cooperation are judged to be morally good everywhere and form the bedrock of a universal moral compass. Those seven principles are: help your kin, be loyal to your group, reciprocate favours, be courageous, defer to superiors, share things fairly, and respect other people's property.[38]

This new idea was quite a big deal because up until then it seemed quite reasonable to assert – as cultural relativists have always done – that there are no moral universals, and each society has therefore had to come up with its own unique moral compass. As I will explain, this is not the case. Moreover, the same seven principles of cooperation on which these moral ideas are based are found in a wide range of social species and are not unique to human beings.[39] These moral intuitions evolved because of their benefits for survival and reproduction. Genetic mutations favouring cooperative behaviours in the ancestors of social species, such as humans, conferred a reproductive advantage on the organisms adopting them, with the result that more copies of those genes survived and spread in ensuing generations. Take the principle that we should care for (and avoid harm to) members of our family. This moral imperative likely evolved via the mechanism of 'kin selection', which ensures that we behave in ways that increase the chances of our genes being passed on by endeavouring to help our close genetic relatives to stay alive and produce offspring. Loyalty to group, on the other hand, evolves in social species that do better when acting in a coordinated way rather than independently. Reciprocity (the idea that I'll scratch your back if you scratch mine) leads to benefits that selfish action alone cannot accomplish. And deference to superiors is another way of staying alive, in this case by allocating positions of dominance or submission in a coordinated fashion rather than both parties fighting to the death.

The theory of 'morality as cooperation' proposes that these seven principles of cooperation together comprise the essence of moral thinking everywhere.[40] Ultimately, every human action that prompts a moral judgement can be directly traced to a transgression against one or more of these cooperative principles. At least, that was the theory. But how could we possibly establish that these seven principles were indeed universal?

The answer lies in an unprecedented study of humans' moral reasoning around the world.[41] My colleagues and I assembled a sample of sixty societies that had been extensively studied by anthropologists and therefore provided rich data on prevailing moral norms in those cultural groups. To qualify for inclusion, each society had to have been the subject of at least 1,200 pages of descriptive data pertaining to its cultural system. It must also have been studied by at least one professionally trained anthropologist based on at least one year of immersive fieldwork utilizing a working knowledge of the language used locally. The sample of societies was selected to maximize diversity and minimize the likelihood that cultural groups had adopted their moral beliefs from one another. They were drawn from six major world regions: Sub-Saharan Africa, Circum-Mediterranean, East Eurasia, Insular Pacific, North America, and South America.[42]

The aim was to mine 400 documents describing the cultures of these sixty societies to establish whether or not the seven principles of cooperation were judged morally good or not, whenever they were mentioned as salient. We found 3,460 paragraphs of text that touched upon these cooperative principles. In each case, we wanted to know whether the type of cooperation described was characterized using words such as good, ethical, moral, right, virtuous, obligatory, dutiful, normative, or any other morally salient language. This produced 962 observed moral judgements of the seven types of cooperative behaviour. In 961 of those instances (99.9 per cent of all cases), the cooperative behaviour was judged morally good. The only exception was on a remote island in Micronesia where stealing openly (rather than covertly) from others was morally endorsed. In this unusual case, however, it seemed to be because this type of stealing involved the (courageous) assertion of social dominance. So, even though this one instance seemed to contradict the rule that you should respect other people's property, it did so by prioritizing the alternative cooperative principle of bravery. The main take-home here is that the seven cooperative principles appear to be judged morally good *everywhere*.

Does this necessarily have anything to do with religiosity, though? After all, there is no obvious reason to think we must have gods to believe in the seven moral principles. However, since so many people in the world seem to think there is a link between being religious and being good, this question clearly deserved close attention.

The answer turned out to be a little complicated. One element of the universal moral repertoire does seem to be intuitively connected with our religious instincts: one that takes us back to the early developing expectation that supernatural agents will be socially dominant. In the preceding section I described our research with babies, showing that even before they could talk, they expected agents with supernatural powers to win out in a power struggle with an agent lacking such powers. This suggests that our relationship with the spirit world is underwritten by a *moral* concern with deference to authority. The gods and ancestors will tend to be our masters and we will tend to be their servants: we will bow down to them and not they to us.

But what about the other moral domains – could they also be linked to our religious beliefs and behaviours? Are our religious intuitions – for example relating to supernatural beings and forces, life after death, or intelligent design in nature – linked to the seven moral rules we have also found to be universal?

The answer is yes – but in ways that are altogether less natural and intuitive. That is, the link between most of the features of intuitive religiosity and the universal moral repertoire is not itself natural but a product of the way religions as cultural systems have evolved as part of our unnatural history of civilization. As such, the links between intuitive morality and intuitive religion have changed dramatically over the course of human history. For example, as we will discover in later chapters of this book, our natural moral imperative to care for kin has been developed in many of the world's religions into an obligation to take care of our ancestors, by dutifully carrying out a variety of behaviours expressing filial piety. The moral rule that we should reciprocate favours also features prominently in the logic of our interactions with the spirit world – and took a dark turn during the phase of religious history when human sacrifice became rampant. Much later still in the history of religions, we observe the emergence of beliefs in gods demanding that humans everywhere obey law-like commandments. Although both religion and morality are rooted in human nature, our intuitions about the two are not necessarily linked. Sometimes, our moral reasoning seeps into our religious beliefs; at other times, it does no such thing. It depends on the cultural tradition you happen to live in.

And so, we have at least a provisional answer to the question: do we need religion to be moral? The answer is no. Our intuitive religious beliefs as well as the many 'catchy' MCI concepts that people adopt would seem to have no necessary connection to our moral intuitions – barring perhaps the special case of deference to supernatural beings. It seems that we can answer Socrates at last. Goodness is not good because it is loved by the gods – rather, it is good whether or not the gods know or care. Unfortunately, this also suggests that most of the people surveyed by Pew on this topic have been misled, however unintentionally, by the religious leaders of the world.

My wife and I have a narrowboat on the River Thames, moored deep in the Oxfordshire countryside. I have long loved to photograph the wildlife from the window of our floating hide. I have witnessed many beautiful scenes over the years – from the arrival of summer migrants at dawn to the lightning strikes of kingfishers and the ghostly glide of the barn owl on a misty autumn evening. But of all the beautiful creatures of the river, there was one iconic species that had always eluded my telephoto lens: the Eurasian otter. I would frequently ask lock keepers for news of otter visitations and then take up position at the locations they considered most likely to produce a sighting. But to no avail.

After one particularly disappointing evening of sitting on the edge of a weir as darkness fell, I made the trek back upstream to our boat and passed a fitful night, dreaming of otters. I awoke with a start around 5 a.m. I could not say what caused it. But for some reason, I crept out of bed and opened a hatch on the side of the boat. Staring up at me from the river was a pair of milky eyes set into a spiky head of fur, literally just an arm's reach away. The otter dived under the boat and emerged with a fish. Then it was off again, swimming towards the stern. I crept out onto the deck with my camera, which lay on a worktop in the cabin. Within seconds I was getting close-up pictures of the otter posing for me in the early morning mist.

So, why mention this in a chapter about religion? The answer lies in the way I reacted to this piece of good fortune. I found myself wondering why I had awoken with a start. And, having awoken, why did I decide to open the hatch and look out? And why at that very particular moment? Had it been a minute earlier or later, there would have been nothing to see. Why was the camera set up and ready to shoot? Normally, I would have removed my telephoto lens and packed everything away after the failed trip the previous evening. All these coincidences seemed pregnant with meaning. People in some cultural groups might say that a spirit had summoned me from my slumber, and there might be a plethora of culturally distinctive glosses on the meaning of this – perhaps some ancestor, or maybe the spirit of the river, wanted to communicate with me. Because I had been raised in a broadly Christian culture, there was perhaps some vague but lurking intuition that God had taken pity on me for all my failed efforts to see otters in the past and decided to reward me with one. At any rate, I felt an inexpressible sense of gratitude but with no clear idea whom to thank: God, the ancestors, or perhaps the legendary Father Thames? These half-formed intuitions rattled around in the back of my mind, even if I shrugged them off as they broke through into fully conscious awareness.

What these intuitive, partially implicit, and admittedly incoherent thoughts reveal is a set of intuitions about supernatural agency and moral reasoning – basic ingredients of religion in the wild. The mystery was not that there was an otter in the river but that I happened to be in the right place at the right time, with the camera set up to take pictures. This seemed to require an explanation and my immediate thoughts appealed to disembodied agency – a supernatural agent who should be acknowledged and thanked. A similar reaction is evident on the face of every professional footballer whose face turns to the heavens in thanks after taking a successful penalty kick.

The same logic is invoked, of course, when things go wrong. The famous anthropologist Sir Edward Evans-Pritchard observed that the Azande of South Sudan attribute all misfortunes (including virtually every human death) to witchcraft.[43] Even what appears to be an accident is typically explained in these supernatural terms. For example, if people are sitting in the shade of a granary and it collapses, causing their untimely deaths, everyone will blame it on witchcraft. This does not contradict perfectly natural explanations for the tragedy. As Evans-Pritchard was quick to point out, the Azande are the first to admit that the granary supports were weakened by termite infestation, and this is why the structure gave way. But why did it happen to fall at the precise moment that people were seated in its shade, resulting in their deaths? For that, the Azande say, the only possible explanation is witchcraft. Although different cultural groups entertain a wide variety of supernatural explanations for both good and bad fortune, we are alike in spontaneously seeking something more than just a dull and prosaic explanation for events that matter to us.

My experience with the otter made me reflect on quite how deep-rooted our tendency towards religion is. As we've seen in this chapter, there are a plethora of implicit beliefs underlying religion everywhere, deriving from panhuman intuitions about mind–body dualism, essentialism, promiscuous teleology, and suchlike as well as concepts that violate core knowledge in attention-grabbing ways. Anthropologists have described the belief clusters that draw on these panhuman intuitive foundations as 'wild religion' because they spread naturally without any need for special cultivation or institutional support.[44] The study of wild religions has led to new insights into why humans defer to religious authorities, such as evidence that even preverbal infants expect those who channel supernatural powers, by defying principles of intuitive physics, to be socially dominant. And we feel a natural sense of obligation to others, rooted in universal moral intuitions concerning family, loyalty, reciprocity, courage, deference, fairness, and ownership. Together, the biases described in this chapter are key to understanding why people around the world tend to subscribe to unfalsifiable and seemingly improbable systems of belief and also why they cooperate.

But there is another aspect to human nature that often gets tangled up with religion, just as with a wide range of other types of human belief systems. It is an aspect that is responsible for some of our most impressive cooperative achievements but also our cruellest forms of oppression and intergroup violence: tribalism.

3.
Social Glue

The Arab Spring was turning to winter. It was late 2011 and Muammar Gaddafi, Libya's dictator, had been captured and killed. Gunshots still rang out but now they were being directed skywards in jubilation rather than to maim or kill. During the revolution, many of those who opened fire on Gaddafi's forces had no military experience. Most had never even held a gun before, let alone engaged in bloody combat with a well-equipped modern army. And yet all these civilians-turned-fighters risked life and limb in the fight, and many were killed. What motivated their acts of self-sacrifice?

This is a question I had pondered ever since living in Papua New Guinea and learning about the daring raids conducted by tribal warriors in the days before colonization and the emergence of the Kivung. What I had been lacking was an opportunity to study extreme self-sacrifice directly. Then the Arab Spring arrived and much of North Africa and the Middle East was convulsed in civil war. One of my students, Brian McQuinn, had spent most of the Libyan revolution in Misrata – a city under siege for many months – and the evidence he had been amassing was extraordinarily detailed and insightful.[1] After a great deal of paperwork and string-pulling, I managed to get a one-way ticket from a kiosk at Istanbul airport that would finally enable me to join him.

As I stepped onto the tarmac amid a swarm of passengers and livestock, I noticed that the terminal building was peppered with bullet holes and the leaves of palm trees were torn to shreds. My fellow passengers were climbing over one another to claim their luggage. McQuinn retrieved me from the scrum and led me to his car. I noticed a hole in the passenger door and picked up a dead shell from beside my feet as we drove away.

McQuinn avoided caffeine, knowing he had a tendency to talk too fast. But as he bustled me into his car, speaking nonstop, I surmised he had drunk several cups of coffee that morning. He had arranged a meeting with Salim Juha, a former colonel in Gaddafi's army who defected on the first day of the revolution. It was the perfect way to get our work started, McQuinn told me enthusiastically. His voice grew distant as I took in fire-damaged buildings and endless rows of burned-out tanks. After collecting our translator, Mohammed, from a street corner, we proceeded to a large mansion with bollards at the gate. McQuinn spoke briefly to the guards, each strapped up with an ammo belt and carrying an automatic weapon. They hauled the roadblocks to one side. As we drove through the perimeter wall, McQuinn turned to me: 'Mr Juha is very powerful,' he said. 'Please don't insult him or we're both in a lot of trouble.'

The mansion was once an oil investment corporation but now served as a base for Misrata's revolutionary forces. Young men with rifles were sprawled on sofas, smoking and occasionally laughing loudly. As the three of us passed, they fell silent, following us with their eyes. We found Mr Juha in the boardroom seated at a long table, capable of accommodating perhaps twenty people comfortably. To his left and right were leaders of revolutionary brigades, wreathed in cigarette smoke. McQuinn, Mohammed, and I were shown to our seats at the table on the opposite side. Mr Juha stared fixedly at me and delivered a formal speech in Arabic. 'We would like to thank you for sending us your son, Brian,' Mohammed translated.

McQuinn was not my son. He was my doctoral student and there were only a few years between us in age. But I decided it was best not to argue. In any case, I told myself, kin psychology was one of the things we were here to study. Before I could make any notes, however, Mr Juha's speech was over, and it was my turn. I hastily mumbled something about my gratitude for the hospitality shown to my son and then asked if he and his colleagues would be kind enough to answer a few questions of anthropological interest. Mr Juha's stare intensified. 'First,' he said, 'I have a question for you. When news broke around the world about the revolution in Libya, had you ever heard of . . . Misrata?'

It felt like a test. I recalled what McQuinn said about not annoying him. Then I remembered one of my most distinguished predecessors as head of the anthropology department in Oxford. It was none other than Sir Edward Evans-Pritchard – the anthropologist we encountered in the last chapter with an interest in witchcraft beliefs among the Azande. During the Second World War, E-P – as he was known to his colleagues – served as an officer for British military intelligence in Libya, and via his voluminous writings I had learned about a war hero in Misrata. That man, it turned out, was Mr Juha's grandfather. The atmosphere brightened a little upon this discovery.

Mr Juha allowed me to explain more fully the purpose of my visit and the questions I wanted to ask him. I told him I was here because I wanted to know what motivated so many ordinary civilians to take up arms in a popular uprising in which death was far likelier than victory. I pointed out that the vast majority of Libyans who joined the insurgency in 2011 had no military training or preparation. What gave them the hope they could overthrow a state-of-the-art, well-equipped army?

Mr Juha began his answer with a comparison to his time in Gaddafi's army. When he had been a military colonel, he explained, it had taken him three months to train new recruits to use a weapon. But the people he had trained here in Misrata could be taught in three hours. They learned not only how to use a gun but also how to fix it when it broke and how to modify it to suit new situations. They had to figure it out for themselves because there were no maintenance depots. People used their initiative because their focus was on winning, not on earning a salary.

This motivation made all the difference. The secret of the revolutionaries' success, Mr Juha told me, was the will to fight. 'When the revolution began there was no compulsion to join. We just called our friends and asked them: do you want to die or not? If you want to die, come with us. If not, go home and stay out of harm's way. This is not a time for reflection and discussion. He who wishes to die can accomplish anything. He who does not should leave in peace.' As a result of this winnowing process, he told me, any one of the civilians-turned-fighters in Misrata's *kata'ib* (battalions) was worth ten times a professional soldier in any conventional army.

Although some of the details of the Libyan context were unique, the willingness of revolutionaries to fight and die for one another was not. For thousands of years, warriors have been prepared to commit acts of astonishing self-sacrifice on the battlefield that are not easy to explain from an evolutionary perspective. After all, there is no surer way to end one's lineage than to die before reproducing. From an evolutionary standpoint, organisms should normally prioritize their own survival over that of other individuals. But there are at least some scenarios in which it would make sense for self-sacrificial behaviour to evolve. Some of these scenarios help to explain the willingness of both New Guinea warriors and Libyan revolutionaries to fight and die for each other.

The evolution of self-sacrificial behaviour

One of the most obvious evolutionary explanations for self-sacrificial behaviour would be that the behaviour increases the reproductive success of close relatives, even if it might imperil the survival of any given individual.[2] Birds nesting on the ground, for example, are extremely vulnerable to attack from aerial predators. To protect their precious eggs and fluffy fledglings they will often attempt to distract attackers. A common ploy is to feign a broken wing, dragging the seemingly lifeless appendage as they hobble away, enabling their young to run for cover. Some species go for a simpler method – screeching and squawking to attract attention. The wiliest of them pretend to be another species altogether, ideally one that is delectable to predators. My favourite example of this is the distraction display of a tiny Australasian bird known as the superb fairywren. True to its name, the male of the species has a splendid iridescent blue crown, cheek, and upper back painted into a raven-black frame; it is easy to imagine why European settlers likened them to fairies. When their young are threatened, they will adopt a strange behaviour on the ground, creeping around in peculiar darting movements and squeaking in a way that – to a hungry predator – appears exactly like a succulent mouse.[3] Some social mammals undertake similarly impressive acts of heroism for their brothers and sisters as well as for their own offspring. For example, in wolf packs and prides of lions, individuals will also fight bravely to protect their siblings.

All these behaviours evolved because they increase the chances that the hero's genes will be passed on, if not by their own reproductive efforts, then through the mating success of their closest relatives. According to the American-born Russian physicist-turned-theoretical-biologist Sergey Gavrilets, however, there is another possible way in which self-sacrificial behaviour could evolve – one that doesn't require such close genetic bonds. In 2014, Gavrilets came to give a talk at my department in Oxford in which he described the evolution of a genetic tendency to cooperate in situations of conflict, both within and between groups. His core idea was that cooperation among unrelated individuals could be conditioned on shared past *experiences* rather than genetic kinship.

This idea excited me because I had already developed a theory that shared experiences played a central role in the psychology of violent self-sacrifice, based on my observations in New Guinea, Libya, and other regions of the world – a theory that we will get onto in a moment. But the piece of the puzzle still missing was how that psychology evolved in the first place. After his lecture, Gavrilets and I continued talking and our conversation continued over the months that followed. This led to the development of an 'agent-based model' that could help to explain the evolution of extreme self-sacrifice among warrior tribes facing daunting threats to their survival.[4]

Agent-based models are computer simulations of how various 'agents' – in this case humans – might behave in a world where assorted parameters – such as resource availability, competition, or even weather conditions – can vary. Such models can help us to explore how different variables and their interactions lead to a range of observable outcomes.[5] Our model was set up in a virtual space, equivalent to a region of the rainforest in Papua New Guinea, populated by many distinct groups, each comprising a population that could bear offspring and thereby increase in size or suffer mortal losses resulting in a reduction in the size of the group. Each group in this landscape faced two kinds of external threats: enemy groups (e.g., raiding by hostile neighbours) and environmental challenges such as natural disaster (e.g., earthquakes, pestilence, malaria). Both threats could result in the untimely deaths of group members, depleting the population and perhaps even causing the group to die out altogether. Crucially, however, the two kinds of threats were not identical from an evolutionary perspective. Any loss in warfare counted as a gain to rival groups in the imaginary landscape, because it increased their relative strength. On the other hand, losses caused by natural disasters did not automatically benefit rival groups (or at least, not to the same extent).

All the individuals in this model were assigned a particular way of behaving in these conditions. The first group were assigned a pair of 'genes' that made them more likely to cooperate when things were going well (success in war or a favourable environment). The second group were assigned genes that made them more likely to cooperate after a negative group experience (loss in war or a natural disaster). Gavrilets ran many simulations of this model to see which groups became more cooperative, and which did not, in the face of randomly changing fortunes. The results were striking. Some groups inevitably suffered more fitness-reducing events – experiences that negatively impacted their ability to survive and pass on their genes. And these were the groups that ended up being more cooperative than those that suffered less. What's more, these positive effects on cooperativeness were stronger when shared suffering resulted from intergroup conflict than from natural disaster. In short, doing badly made the group pull together more tightly.

This model helped to explain why heroism in battle – despite its deadly consequences for the hero – could have become embedded in our evolved psychology. It showed how shared suffering came to be a potent bonding agent, increasing our willingness to defend the group, whatever the cost.

Ritual ordeals and shared suffering

Unpacking the psychological mechanisms involved in this process has taken many years and began well before I met Sergey Gavrilets. Indeed, my fascination with this problem first began back in the 1980s when I was initiated into the Baining tribe, deep in the rainforest of Papua New Guinea.

You cannot be regarded as a fully grown adult among the Baining if you have not been initiated. Among other things, you would not be allowed to wear the sacred costumes used in dances. It was therefore a moment of great pride for me when I first danced as an initiated man shrouded in foliage and masked by a bark-cloth face covering that together would have made me unidentifiable were it not for the white skin on the backs of my hands. This outfit was traditionally worn when carrying out acts of homicidal violence in a trance-like state, under the cloak of anonymity. In the past, the right to wear such costumes had been acquired by undergoing agonizing procedures. This included the insertion of sharpened bone into the skin at the base of the spine, which was then used as an anchor point to carry the weight of a heavy mask during hours of dancing. The bright coloration on the headdresses was meanwhile produced from the blood of their wearers by scraping their tongues with a sharpened leaf and then spitting repeatedly until the costumes glistened crimson.

Although I danced and sang proudly with my fellow initiates, young people like me were spared the most painful elements of the initiation nowadays due to a prohibition imposed by the Kivung's leaders on the killing or harming of others. Thankfully, this meant I was never ritually mutilated or required to join a raiding party against the group's enemies. During the process of induction into the secrets of the male cult, however, I was provided with vivid descriptions of the ordeals of older men and left in no doubt about the traumatic nature of the process. Men who had suffered together in this way became warriors. And true warriors could depend on each other to cover each other's backs, not only during daring raids but when their communities came under attack.

Humans have long understood that shared suffering creates extremely powerful bonds capable of motivating acts of heroism in battle. For millennia, warrior groups have exploited this fact by requiring their recruits to go through traumatic initiations. In the Sepik River region of Papua New Guinea, Kanigara men endure extensive cutting of their skin, resulting in permanent scars.[6] In other parts of Papua New Guinea, boys and unmarried men are forced to undergo similarly horrendous torments, ranging from piercing of the nasal septum to insertion of barbed plant matter into the urethra to create an arc of blood as it is torn out of the penis.[7] And this is not unique to Melanesia by any means. Across the Arafura and Coral Seas on the landmass of Australia, indigenous groups developed excruciatingly painful forms of circumcision that completely transformed the appearance of the penis. Not all such rituals focus on genital mutilation, however. Still further away across the Pacific Ocean, Plains cultures of North America performed Sun Dances on the end of ropes attached to their bodies by hooks buried deep under the skin. Meanwhile, Amazonian initiates even today put on gloves filled with angry bullet ants, so named because the sting of a single ant is as painful as a bullet wound and lasts for hours. Examples like these could be multiplied indefinitely.[8]

Anthropologists have shown that the most warlike societies in the world also have the most painful initiations, typically leaving permanent scars that bear witness to their sufferings.[9] A common function of these agonizing ordeals seems to be to provide military cohesion. The bonds created through the sharing of painful rituals make participants more willing to support each other when they are in trouble and to fight and die for their fellow initiates if necessary. We have found evidence for this all around the world. For example, building on my studies of initiation in Papua New Guinea, my colleagues and I turned our attention to the effects of similar ritual ordeals in West Africa, where violent clashes over access to land have become an increasingly common feature of interactions between herders and farmers. Population pressures are creating intense competition for access to land and water, and the arising tensions are exacerbated by climate change. In some parts of Africa (especially in countries like Mali and Nigeria), these divisions are regularly exploited by militant Islamist groups. But what causes local conflicts to spiral out of control seems to be less the material conditions of any given region, and more the kinds of ties that connect people to their tribal identities and ancestral lands – and, in particular, the role of initiation rites in group bonding.

This, at least, was the finding of a study that my colleagues and I carried out with 398 participants from the Adamawa region of Cameroon, near the border with Nigeria – a region where thousands have been killed in these kinds of clashes, and many more driven out of the area.[10] Our research found that the willingness of farmers to fight and die for each other when conflicts like this flare up was strongly related to shared experiences of terrifying initiation rites in childhood. As young boys they were taken to a sacred enclosure near the chief's house and menaced by a masked figure who emerged from the shadows as they entered. The boys would attempt to flee in terror, but their initiators would block their exit. Nearly hysterical with fear, this was an experience never to be forgotten. Although they were not physically injured (except for any self-inflicted wounds from their efforts to escape the compound), the psychological impact was permanent.

Extreme initiation rites are not unique to warriors in indigenous or small-scale societies, however. For example, between 1997 and

2006, the far-right drug-trafficking group known as the AUC in Colombia typically required new recruits to carry out ritualized acts of torture, dismemberment, and murder of captives, expecting them to carry severed body parts with them at all times.[11] Similar tactics were used by the Revolutionary United Front in Sierra Leone, who required new recruits to participate in ritualized murders, often including the killing of family or community members.[12] Even conventional armies may inflict physical and psychological tortures on new recruits as part of their training. For example, soldiers who sign up for Taiwan's Amphibious Reconnaissance Unit must undergo a series of extreme ordeals, such as being required to crawl semi-naked over fifty metres of sharp coral while performing compulsory exercises along the way that maximize the pain inflicted.[13] Salty water is meanwhile thrown onto the open wounds of participants to increase the pain. Participants who make errors or cut corners are sent back to the starting line to begin the agonizing ordeal all over again.

Thanks to all these different sources of evidence and many more, the link between rituals involving shared suffering, group bonding, and heroism in battle is long established. But how exactly does this work psychologically? What is it about painful ordeals that creates such powerful bonds that people are willing to fight and die for each other? The answer seems to lie in a remarkable form of group alignment known as identity fusion.

Shared essence and fusion

Identity fusion is a visceral feeling of oneness with the group.[14] It occurs when your personal identity becomes 'fused' with a group identity. If you are a serious football fan, you will probably know what this means. Think about the stadium of the club you love. Then imagine drunken supporters of a rival team vandalizing it (be sure to conjure up a scenario that is both spiteful and gratuitous). Would you take it personally? Would it be as if someone had desecrated your own home? If yes, you are most likely fused with your club.

Not everyone is fused with a football club, obviously. Some are fused with a religious group. Others are fused with their nation. Still others with an ethnic group. Some people even fuse with the choir they sing in or with the posse they go clubbing with. But even if you haven't fused with any of these kinds of groups, you might at least be fused with your family. Again, imagine if a family member was grossly insulted, injured, or kidnapped. How far would you go to defend or rescue them? Would you pretty much stop at nothing? If yes, then you are probably fused with your family.

Fused individuals are passionate defenders of the group – not only willing to cooperate more with other group members but even to make the ultimate sacrifice to protect the group when it comes under threat.[15] Obviously, this is exactly the kind of attitude you would want in a military group – especially in times of war. Sure enough, fusion is exploited by a wide range of violent groups, from street gangs to terrorist cells and from conventional armies to suicide bombers.

The fusion construct was first developed by a team of social psychologists from Texas and Madrid, led by William B. Swann and Ángel Gómez respectively.[16] They initially based the construct on a measure of 'inclusion of other in self', using circles to represent 'self' and 'other' and the degree of overlap between them to indicate how much you consider the other to be included in your concept of self. What Swann and Gómez realized was that a similar pictorial measure could be used to ascertain the extent to which people consider their personal identity to be encompassed by that of the group. They drew up a series of images showing degrees of overlap between a little circle and a big circle. And they told research participants that the small circle is 'you' and the big circle is 'your group'. Which one of the following pictures, they asked, best captures your relationship with the group? Those choosing the image on the right of the sequence, where the small circle was completely encompassed by the big circle, were said to be fused.

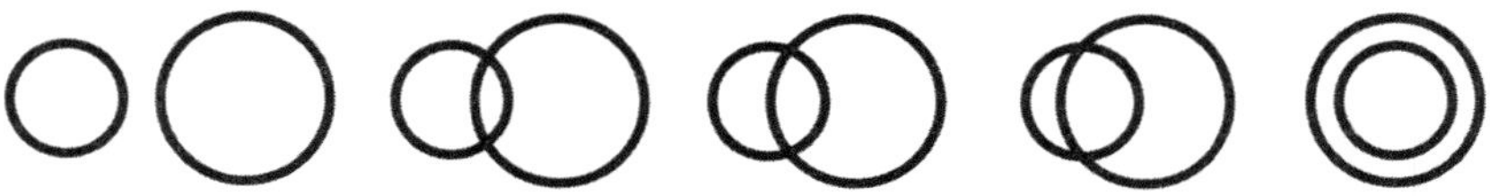

Fused

Adapted from Swann, W. B., Jr, Gómez, Á., Seyle, D. C., Morales, J. F., & Huici, C., 'Identity Fusion: The Interplay of Personal and Social Identities in Extreme Group Behaviour', *Journal of Personality and Social Psychology*, Vol. 96, No. 5, pp. 995–1011 (2009)

When I first came upon this fusion measure, it immediately struck a chord. For years I had been trying to understand why painful initiations in Papua New Guinea created such strong bonds between those undergoing them together. I felt sure it was linked to the way memories for those experiences shaped one's personal identity and also the way such memories influence the way we think about the groups involved.[17] But what I lacked was a way of measuring the effects of these processes on group bonding. The Swann–Gómez fusion construct seemed to be exactly what I'd been looking for.

After a great deal of research using the new fusion construct, my collaborators and I arrived at a testable theory of how the mechanism works. According to this theory, there are at least two pathways to becoming fused.[18] The first one involves the sharing of biological essence with other group members. Across a great diversity of human societies, it is common to imagine that members of the same family share heritable essences – variously construed as shared blood, bones, spirit substance, and suchlike. Does believing that we share essentialized biological traits like these make us more 'fused'? One of the early ways we went about testing this hypothesis was by running a study with twins. We recruited 246 non-identical twins and 260 identical twins and asked each pair how 'fused' they felt with one another. Unsurprisingly, perhaps, the identical twins said they were significantly more fused than the non-identical twins.

But even when there is no actual biological basis of fusion, humans are intriguingly prone to invoking biology to explain their group bonds. Indeed, even just the *perception* of shared biology can in some cases be used as a basis for fusing with large groups that share physical traits relevant to racial categorization – such as skin colour, hair type, or facial characteristics. The power we ascribe to biological bonds would also help to explain why people who are fused often talk about other group members using the language of kinship. Take a nationalist describing their country as a motherland or fatherland, men of the same ethnic group describing each other as brothers, feminists construing womankind as a sisterhood, and so on.

It is plausible that our willingness to make sacrifices for the group evolved via biological kinship, just like the heroic behaviour of the superb fairywren. But we were also interested in how fusion came about between people who shared no biological bonds – like those soldiers 'fused' with a military unit, for example. In line with this, our twins study showed that independently of biological relatedness (which has long been known to drive heroism via kin selection), willingness to sacrifice for your sibling was driven by shared *experiences*. This supported our hunch that there were at least two pathways to fusion – not only shared biological ties but also the feeling of having been through something personally transformative with other members of a group.

This is where those non-genetic bonds – your football club, your church, your country – come into play. Your bonds with such groups are primarily forged not by blood ties (real or perceived) but by specific memories and moments. Maybe you remember going to a rock concert with your best friends as a teenager and feeling moved by the lyrics of the songs and connected to each other in a way you will never forget. Or maybe you all risked being arrested on a protest march. Or perhaps you suffered together at a sports event when the team you all loved endured a humiliating defeat at a crucial cup final. If you and your buddies have ever been through an experience of this kind, you will probably recognize the shared-experiences pathway to fusion. These are the kinds of experiences that link us so closely to a club, to a band, to a bunch of friends that we feel we could follow them loyally to the ends of the earth.

I dubbed this route to group bonding the 'imagistic' pathway, because it revolves around mental images associated with past events that have a particularly haunting and meaningful quality.[19] Bonds born not of blood ties, but of powerful images marked indelibly on our memories.

The origins of fusion

What is going on in our heads when we become fused to a group? Part of the answer lies in the way we remember particularly life-changing experiences and reflect on their meanings and consequences. In the case of our memories for very emotional life events, we are inclined to invest a lot of effort reflecting on why things happened a certain way, what significance they may have had then and subsequently, how differently things might have turned out if those events had never happened, and so on. It has long been recognized that these processes feature prominently in the way people think about ritual ordeals. For example, the famous Norwegian anthropologist Fredrik Barth argued that the agonizing initiations of Inner New Guinea caused participants not only to remember their experiences for many years to come but also to engage in prolonged reflection on the symbolism and significance of those events, creating what he called 'fans of connotations' linking together the memories and emotions and exploring the analogies between various images they conjured up.

Building on this idea, I came to refer to this process as the 'imagistic mode of religiosity' – mainly because of the focus on potent imagery and its interpretation. I found evidence for this not only in New Guinea initiation cults but in many other kinds of groups performing transformative rituals all around the world and over time – from the Dreamtime visions of Australian aborigines to the spirit journeys of Amazonian shamans. At the heart of this imagistic process was usually an emotionally intense experience which endured in memory and demanded interpretation. I came to see this as the foundation on which fusion with unrelated group members was based. But to understand how it worked, we needed to design some experiments.

Our first objective was to ascertain whether there was indeed a link between the emotional intensity of group rituals and the investment of meaning in those experiences. Building on this, our next objective was to establish whether these kinds of imagistic experiences did indeed lead to identity fusion – and if so, how that process worked.

Psychological experiments are typically carried out in laboratories – rooms on university campuses that have been set aside for that purpose. However, my colleagues and I wanted to run controlled experiments on the effects of participating in rituals that more closely resembled the painful initiations of warrior groups. The first artificial rituals we created were conducted outdoors, in a wooded field accessible from Queen's University, Belfast, where I was working at the time. We set out to see what effect ritual-style behaviour had on the way our study participants processed their memories.

The rituals we designed were loosely based on those I had studied in Papua New Guinea, even if the specific objects and environments were inevitably quite different from those I experienced in the tropical rainforest.[20] Students participating in these rituals were formed into groups of about fifteen participants – around the size of a typical initiation grade in some New Guinea tribes. On arrival, they were told we would be performing a propagation ritual learned from a distant tribal group. The purpose of the ritual, we further elaborated, was to promote the fertility of the natural world, for example to make more abundant the animals and wild foods that the group depended on. Other than that, however, we said that little was known about the meaning of the ritual or whether it was actually effective in promoting fertility. Participants were then led in a solemn procession across the field, to a location in the surrounding woodland where some tree stumps had been arranged in a circle, in the centre of which some rotting leaves had been piled up. The students were told to 'wash' their hands in the rotting leaves. Then they were given cloaks to wear and led to another location, and each given a long stick. The students were now told to break off the end of the stick and throw it away behind them, and they were then given a stone and instructed to grind it into the earth. Next, the experimenter told them to 'plant their spears' (indicating what remained of the sticks) into the earth between their feet. Finally, one by one, each of the participants was led to a hole in the ground, marked by a torch that burned with a living flame.

As each participant approached the hole, a drum was beaten rhythmically. Then, as they knelt down at the site of the hole, the drumming stopped, and the participant was invited to place one hand into the hole while a string of magic spells was uttered in the language of the people I lived with in Papua New Guinea (and which we could be sure that none of the participants could understand). Now, if you have ever been asked to place your hand into a hole without knowing what lies inside, then you will know this isn't an easy task. Many grimaced as they did so, withdrawing their hand in fright before trying again. Having completed this ritual, all participants were instructed to hurl their sticks into the woodland with as much force as they could muster and then to return to the exit point of the field in single file.

Everyone in the study followed the exact same procedures described above. But since we were interested in the effects of emotional intensity, we assigned each student to one of two contrasting conditions: a 'low-arousal' and a 'high-arousal' condition. In the low-arousal condition, the ritual was performed in the middle of the day, in a relaxed and lowkey way, minimizing any sense of drama or excitement. In the high-arousal condition, by contrast, the ritual was performed in the evening, as the light was fading. Furthermore, during the phase when participants approached the hole in the ground to the accompaniment of a drumbeat, those assigned to the high-arousal condition were required to wear blindfolds.

After performing the ritual, we measured how emotionally intense the experience had been for participants using a self-report scale. We were particularly interested in whether differing levels of fear during the ritual would affect how deeply participants reflected on the meaning of the experience during the weeks that followed. We measured this by conducting questionnaires at successive time points up to two months after participants had undergone the ritual experience. Each participant was asked to recall what happened in the ritual sequence and for each step they were asked what they thought it meant. Later, we analysed their statements to capture two aspects of interpretive richness. One focused simply on how much reflection was evident in their response – a raw count of the number of actions accorded meanings. The other focused on what we called 'analogical depth' – that is, attributing a specific meaning to the action such as 'grinding the stone into the ground symbolized the planting of a seed'.

We then divided the participants into those who had found the rituals emotionally intense and those who hadn't. To investigate the effects of this, we had the participants describe afterwards what happened during the ritual and to list any thoughts they had experienced about what each element meant. Then we added up how many meanings they had produced (giving extra marks for analogical depth). We found something striking: those who experienced the ritual as more frightening reflected more deeply on what had happened, producing richer symbolic interpretations. This experiment provided the first clear evidence in support of our hypothesis that we reflect more on ritualistic experiences shared with others and therefore find them more meaningful if they are emotionally intense. This would later become the foundation stone for our theory of the imagistic pathway to fusion – our idea that it is by undergoing such experiences that participants in extreme rituals become bonded so tightly together that they will stop at nothing to defend each other.

However, theories in science cannot be declared settled based on a single study. Experimental findings need to be replicated. And each time a replication is attempted, something new is usually learned in the process. In our next experiment, we wanted to introduce a more direct measure of emotional intensity based on objective physiological evidence. In our first study, it was unclear how accurately or consistently participants rated their emotional experience. It is possible that some were misrepresenting the intensity, whether by over- or underrating it. We also wanted to create a more emotionally intense experience for participants. The ritual process we had created was much tamer than the real initiation rituals that young people in Papua New Guinea traditionally underwent. As such, even though we managed to achieve measurable differences in how emotionally intense the experience had been, the high end of the scale was still very low compared with the 'real' thing in a traditional culture. Indeed, it is probably fair to say that riding a big dipper or watching a good horror movie would have been a lot more frightening than going through our artificial ritual.

We spent a lot of time thinking about how we might improve our measures of physiological arousal during ritual performances, allowing us to assess the accuracy and consistency of our self-report measures of emotional intensity. In the end, we decided we needed to measure a combination of heart rate during participation as well as electrical conductivity on the surface of the skin (so-called 'galvanic skin response' or GSR, used in experiments as an objective measure of emotional arousal – especially fear). Spikes in these two physiological measures relative to baseline data collected before the ritual began would provide us with more objective measures of how frightened participants really were during the experience. The only problem was that this involved the use of equipment that around the turn of the millennium, when we were running these studies, was rather big and heavy. They required participants to have wires attached to them. This was not an ideal setup for carrying out rituals in the field.

In the end, we went back to the drawing board and started working on an entirely new design for our artificial ritual. We needed an evocative environment, capable of stimulating the senses in emotionally impactful ways – but which could also accommodate a bulky GSR machine. After much head-scratching, we came up with an ingenious answer: caves. In ancient foraging societies, our ancestors produced astonishingly beautiful works of art on the walls of caves deep underground. Archaeologists have argued that some of the images were deliberately painted on surfaces that could only be viewed as intended by crawling into tight spaces and contorting one's body into awkward positions. Musical instruments and heel prints scattered on the floors of ancient caves suggest the performance of rituals amid the artworks, perhaps conducted by the light of burning torches.[21] Combine that with the strange lighting conditions and acoustics of these eerie spaces and you have the makings of a primitive but highly effective set of special effects. Our goal was to try to do something similar on a modern university campus. But how?

The solution arrived in the form of a facility that my colleagues in Queen's University musicology department had recently created. They called it the Sonic Arts Research Centre (or SARC).[22] It was designed not only as a bookable performance space but as a place to conduct experiments on the effects of music on our psychology and behaviour. There were powerful speakers seemingly everywhere, including under the floor, allowing sound to be projected from all directions. And the walls were constructed of materials that could be moved around to create various alterations to the acoustics of the performance space. Sophisticated stage lighting enabled users to create a diversity of visual effects and changing colours. This was as near as we could get to a Palaeolithic cave system without actually going into one.

The next question was how to create a suitable ritual in these cavernous surroundings. One constraint that became obvious straight away was that, to allow us to wire participants up to heavy equipment, we needed them to perform the ritual one by one rather than in a group. We also needed participants to give up the use of one hand so that we could attach sensors to the tips of both index and middle fingers, in turn connected to the GSR monitor. This meant that the ritual needed to be carried out from a stationary location and any actions limited to the use of one hand.

These constraints at first sounded daunting, but in time we came up with an action plan that was reasonably similar to real-world rituals. In our new iteration, we had participants sit on a chair in the middle of the SARC performing space. In front of them was a table covered with a hessian cloth, which we described as an 'altar'. The altar was laden with various items to be used in the ritual performance: a bowl of oil, a small lump of clay, and a necklace. Beyond the altar was a screen onto which instructions could be projected. All participants in the study were instructed (via the information projected onto the screen in front of them) to perform certain actions in sequence. To begin, they had to dip a finger into the oil and dab it on the forehead and neck. Then they had to recite a chant while drawing an imaginary triangle around the necklace on the altar before placing the necklace over their heads and wearing it. Some music was played and participants were asked to focus intently at a point on the screen. Meanwhile, a man in a fur coat, resembling a large animal, entered the room carrying a small box. He walked up to the altar and placed the box on it. The music stopped. Then the man in the fur coat shook a rattle and left. Following his departure, some drumming was played through speakers. Participants were asked to remove a smooth black stone from the box and then wrap the clay around the stone and place the resulting object back into the box. Then they were asked to remove the necklace and place it into the box as well.

As with the earlier 'propagation ritual', the goal of this new 'altar ritual' was to measure the effects of emotional intensity on how much people subsequently reflected on the meaning of the experience. In that regard, we were helped by the fact that most of the students participating in this experiment would have found the high-tech environment of the sonic arts performance space somewhat bizarre to start with. Add to that the weirdness of being wired up to machines and approached by a strange man covered in fur, shaking a rattle at them, and we could be confident the experiences would be at least somewhat imagistic. Nevertheless, to really create an impression we wanted to go much further than that. This time, in the emotionally intense 'high-arousal' version of the ritual, we started by replacing the houselights with a more menacing red lighting. To increase the sense of drama surrounding the arrival of the man in the fur coat, we had him walk behind each participant while shaking his rattle; they would know he was there but would be unable see him. Also, the box containing the stone was sealed so that participants could not see inside it and had to reach in through a hole in the side of the box without knowing what their fingers might encounter in the process. And to be absolutely sure that the emotional impact would be higher in this condition, the drumming played during the procedure was very loud and gradually increased in tempo, taking full advantage of the acoustic effects that SARC was able to provide, projecting sound from all directions. All these elements were removed or greatly reduced in the low-arousal condition: for example, the lights were turned down rather than turned off, the music was played quietly, the man in the fur coat remained fully in view, and the box was open so that participants could easily see what it contained.

We found that these differences between high- and low-arousal conditions had their intended effect on self-reported levels of emotional intensity. Importantly, though, we now also had more objective physiological measures that confirmed these self-assessments, showing that heart rate and GSR did indeed spike in ways that correlated with people's descriptions of how they felt at various points in the ritual. As before, we sorted people into groups displaying high and low arousal – and then asked them how much they reflected on the meaning of their experiences over the next month. And we used the same methods as before to measure the effects of participation on subsequent reflection and meaning-making. Once again, the findings were startlingly clear: those whose experiences of the altar ritual were the most emotionally intense also reflected more, and reflected more deeply, afterwards.

These findings constituted a major step forward in understanding the effects of participating in imagistic rituals. But as interesting as these results were, they were not, in themselves, the subject I was most interested in studying. Remember, my model posited that emotional intensity was not important in and of itself – but because of the way in which it contributed to *group bonding*. Having shown that emotionally intense experiences led to more self-reflection, we now needed to establish whether this self-reflection was associated with higher levels of cohesion in the group and the willingness to fight and die for other members of the group when threatened. In this context, the idea of identity fusion was a game changer.

As soon as the fusion construct had been developed, we began to design a plethora of new studies that would allow us to measure the impact of shared emotion and meaning-making on cohesion and self-sacrifice for the group. One of our first studies using the new fusion measures sampled around 200 people on both sides of the sectarian divide in Northern Ireland – those who had endured shared negative experiences as a result of the Troubles, dividing Catholics and Protestants in the region.[23] First, we asked participants to consider a list of the most common forms of sectarian abuse (being verbally insulted, publicly humiliated, having their property damaged, etc.) and asked which of these they had personally experienced, as well as how frequently and severely they felt they had suffered – whether physically or emotionally – as a result. As with the earlier artificial ritual experiments, we wanted to know about the effects of distressing experiences on subsequent reflection. So we asked participants how often they had thought about these experiences on a scale of one to six, with answers such as 'I have only thought about this a little bit' at one end to 'I have spent many years reflecting on them' at the other. Finally, we asked each participant how strongly fused they felt with their 'side' of the conflict, whether Catholic Republican or Protestant Unionist.

What we found was that Protestants and Catholics who said they had experienced more distressing episodes as a result of the Troubles also ended up reflecting more on their negative experiences, as measured using the questions just listed. The same was true of those whose suffering had been the most intense. In other words, *all* our measures of shared suffering resulting from sectarian violence were associated with higher levels of reflection. This was very much in line with our previous research. The crunch question, though, was how these findings related to identity fusion. And here too, our hypothesis was vindicated. We found that the more our research participants had reflected on their experiences of shared suffering during the Troubles, the more fused they were with their 'side' in the sectarian conflict. The process of reflection seemed to be playing a crucial role in the journey to identity fusion.

Between them, these experiments added up to a dramatic discovery. They suggested that emotionally intense experiences can change us fundamentally as people and by sharing these self-defining experiences with others, our personal and group identities become fused together. The evidence was growing: memories for shared, emotionally intense experiences do indeed lead to a very powerful form of group bonding.

We knew, however, that this evidence alone was not enough to prove our theory about the 'imagistic' pathway to fusion, rooted in shared memories. We realized it would require many studies replicating these findings across a variety of populations before we could feel really confident that we had nailed it. Our research group therefore went on to explore the relationship between shared suffering, reflection, and identity fusion with many different groups – ranging from soldiers on the frontline[24] to football fans who lose crucial matches[25] and from Muslims radicalized by atrocities in the news[26] to animal lovers appalled by the cruelty of trophy hunting.[27] In one study, one of my doctoral students, Tara Tasuji, set out to explore the effects of one of the most intense experiences of all: childbirth. In a survey of 164 mothers, we showed that fusion with other mothers increased significantly after childbirth.[28] Subsequent rumination on birthing experiences (just like reflecting on the Troubles in Northern Ireland) led to higher levels of fusion with others who had similarly suffered.

As a result, we now have a much better understanding of how

people become fused with their groups. It all boils down to feelings of shared essence. This essence could be rooted in perceived biological substance – for example, blood or bones or (nowadays) genes shared among people with a common ancestor. But, more intriguingly, it could be rooted in a potent memory that is essential to our group identities – such as a traumatic experience that changed our lives and came to define 'us' as a group.

Fighting and dying for the group

Having achieved a fuller understanding of how people fused with groups, the next question was how this affected their behaviour – in particular, their willingness to risk life and limb to protect the group. At first, our ideas about the consequences of identity fusion were based on hypothetical scenarios in which we measured people's willingness to endorse self-sacrificial behaviour. Basically, we were asking people what they would do if they were ever faced with an extreme scenario in which the most effective way to defend others in the group was to lay down their own lives. But what people say they would do and what they would actually do when faced with such dilemmas is not necessarily the same thing. We needed to find a situation where the stakes were for real. This is what led me to Mr Juha, the Libyan rebel leader, in Misrata in 2011.

Once I had convinced Mr Juha that I genuinely did know where Misrata was before the Arab Spring had begun, I asked him for an example of heroism. In a room full of people who had lost scores of their closest friends in frontline combat, this was no doubt a tactless question. But he didn't blink. Mr Juha fixed me with his steely blue eyes through a thick cloud of cigarette smoke, and he asked if I'd heard about the boy on Tripoli Street. I hadn't. Apparently, nobody much outside of Misrata had heard this story – not even the many eager journalists from Al Jazeera.

'The tanks were driving into Tripoli Street and there was a flag on the back of this tank. A kid of maybe fourteen or fifteen, he climbed onto this tank as it was moving,' he said. 'Why? Just to remove the green flag and put the other flag. The revolutionary flag.'

'Just a flag?' I asked.

'Yes, just the flag.'

The boy expected to be killed, Mr Juha explained. And he was.

It is more than likely that the boy was fused with his fellow revolutionaries. Since the fusion construct was first developed, numerous studies have shown that being fused with others predicts willingness to fight and die for the group more readily than any previously established measures of group cohesion, particularly when people's attention is focused on their personal or group identities.[29] Studies have likewise shown that increasing the agency of participants by getting them to engage in forms of exercise that elevate heart rate and other autonomic functions, further amplifies the effects of fusion on willingness to fight and die for the group.[30] This makes sense theoretically because the fusion of personal and group identities implies that amplifying the salience of either one should increase the effects of fusion on behaviour. An obvious limitation of such measures, however, is that it is relatively easy to choose high scores on a scale expressing strength of feeling or willingness to engage in self-sacrificial behaviour in the abstract. Perhaps we would feel differently if confronted with life-and-death decisions in a more concrete way.

To make such decisions feel a bit more lifelike, researchers at first turned to intergroup versions of the famous 'trolley problem'. Trolley problems are thought experiments in which participants are invited to imagine a vehicle on a track (the 'trolley') hurtling towards some disastrous situation, typically one in which innocent people will die unless the participant in the study is willing to take some action to avert it. But there is usually an ethical dilemma. For example, by acting to save some people's lives, others will inevitably be killed.[31] Researchers into identity fusion soon realized that such questions offered a simple way of exploring how far people would be prepared to go to defend the group. For example, some studies showed that the more fused you are, the more willing you would be to jump to your death to prevent a runaway trolley killing fellow group members. Even more strikingly, highly fused study participants were more likely to push aside a fellow group member on the brink of carrying out an act of self-sacrifice to save others. They were prepared to die heroically in their stead, thus saving the life of even more ingroup members.

To some of us, however, these experiments still felt a little too abstract. Indeed, a limitation of many fusion studies is that we cannot know if *endorsement* of self-sacrifice, even in relation to concrete trolley problems, would translate into *actual* fight-and-die behaviours in the real world. This is not to suggest that participants in such studies are insincere; it is just to acknowledge that feeling you would be willing to jump in front of a speeding trolley in theory is very different from actually doing so in practice. Obviously, it is not easy to run empirical studies on such a topic. Apart from anything else, the people who have most convincingly demonstrated their willingness to sacrifice themselves for the group are already dead and therefore unavailable to be interviewed.

This is how I ended up in Misrata, Libya in 2011. There it was possible to recruit large numbers of research participants who had already demonstrated their willingness to die for each other in ways that were impossible to fake. The purpose of my visit was not simply to talk with the leaders of the revolution but to obtain their permission to run a survey with the fighters under their command. Our goal was to measure levels of fusion and try to understand what drove insurgents to lay down their lives for each other.

The result was a survey in which we measured levels of fusion with family, with fellow fighters in one's own battalion, with fighters from other battalions and with supporters of the revolution who did not themselves fight.[32] The findings were more striking than those of any fusion studies we had yet undertaken. For one thing, we encountered by far the highest levels of fusion we had yet recorded – for both family and fellow fighters, fusion scores were at ceiling level across the board. By contrast, the fighters were not fused with mere supporters of the revolution – those who were on the same side ideologically, but who lacked the shared experiences of suffering which fighters and their families had endured in the months preceding the survey. Simply supporting the same side or sharing the same beliefs was clearly not enough to drive fusion. What fused Libyans together in 2011 was the shared experience of coming under fire and fighting back together.

But there was more. We wanted to find out whether the most intense forms of suffering experienced during the revolution would predict the highest levels of fusion and self-sacrifice. This was difficult since all 179 revolutionaries interviewed had suffered acutely, as had their families, and all were as highly fused as it is possible to be using our standard pictorial measure. And so we added an innovation into our approach: a forced choice question. We asked the fighters to decide which of the groups they were most fused with they would choose if they could only nominate one. Were these fighters more fused to their families or their fellow revolutionaries?

This question uncovered a striking difference between two categories of revolutionary. Providers of logistical support within the battalions were more likely to choose their actual brothers over their brothers in arms. By contrast, members of the revolutionary battalions – those who fought on the frontline – were more likely to choose their comrades. They felt more fused with their fellow fighters than with their own flesh and blood.

This was a remarkable finding. We had discovered that 'imagistic' fusion – the kind rooted in shared experience – is not just another way of feeling *like* family. In fact, in periods of intense violence it could be even *stronger* than ties of biological kinship. And in the years since, the evidence has only grown. As a result of this kind of research we now know that fusion helps motivate extreme forms of self-sacrifice around the world – not only for members of our families but also for those who share our most self-defining experiences. My colleagues and I have found the same patterns repeated not only among revolutionaries in Libya, but men in West African villages who have endured terrifying initiations, American soldiers who experienced the horrors of combat in Vietnam and Iraq, and even football fans who have shared the ordeals of defeat in football stadia.[33] We now know also that bonds of brotherhood are linked to willingness to fight and die for others,[34] and that feelings of fusion and kinship together make us more willing to make sacrifices for unrelated group members.[35]

We had come a long way from those first inklings that unpleasant experiences can underpin group bonding. Indeed, we had learned that not only shared biology but also shared experiences can contribute to identity fusion, and in turn motivate a willingness to engage in extreme forms of pro-group action. Seen in this light, it is not really surprising to find that military groups have always exploited both ideas. They create a strong sense that their warriors are somehow biologically related, by emphasizing their physical similarities – making them look as similar as possible and having them march and parade in synchrony. But they also create a sense of 'brotherhood' by putting their warriors through extreme forms of shared experience – not only on the battlefield but through the agonies of initiation, bootcamp, and various forms of ritualized hazing. In the process, they harness one of the most powerful forces in the world to military ends. The imagistic pathway to fusion: creating water that runs thicker than blood.

A telling final example of self-sacrifice is the story of Lieutenant John Fox of the 92nd US Infantry Division. In the closing months of the Second World War in December 1944, Fox volunteered to stay behind as an observer in an Italian village after American troops were forced into retreat by advancing German forces. From his vantage point on the second floor of a house, Fox radioed details of the locations of the German troops on the ground, ordering artillery fire on key targets. One of those targets was his own location. On the receiving end of this order was his close friend, Lieutenant Otis Zachary, who thought that Fox must have given the order by mistake. But Fox insisted they go ahead. The last words Fox uttered before being killed in the barrage were: 'Fire it! There's more of them than there are of us. Give them hell!'[36]

Although we may gaze in astonishment at a fairywren pretending to be a delicious mouse to distract a predator, many of us would probably do much the same thing to defend our offspring. This can be explained quite easily from an evolutionary perspective. What is harder to explain is why we would risk life and limb for genetically unrelated individuals. And yet, like John Fox, humans do this time and again when their backs are against the wall. People commonly associate acts of extreme self-sacrifice with Islamist martyrs in the Middle East, Pakistan, and Southeast Asia. But the phenomenon is actually much older and more widespread than that. Throughout recorded history, people have fought for their groups knowing that they risk dying in the process and in some cases in the certain knowledge that they will die violently. Feelings of kinship often play a role – including the language of 'brotherhood' or 'sisterhood' that everyone knows deep down is just a fiction. But there is another explanation: a feeling that what makes you, you and what makes me, me are the same thing. The sense that we have a shared essence – one rooted in experiences that define us personally, but which are also essential to the group. In this way, we become more than just neighbours. We become a *tribe*.

We've seen in this chapter how the most extreme forms of tribalism, including willingness to fight and die to protect other group members, are associated with an especially intense form of group bonding – one in which personal and group identities become fused together. Our research on fusion has led us to interview not only battle-hardened warriors, but mothers who have suffered traumatic births and the fans of football clubs enduring painful losses. As a result, we can now explain with some confidence when and why people fuse with the group, and how far they will go to protect their group's interests.

As a result, we find ourselves at a very interesting point in the study of human nature. Over the course of the last three chapters I have set out evidence for three natural biases that underpin much of human social behaviour: the tendency to copy our peers, to believe in supernatural forces, and to bond with groups. But all this raises a question. If these biases are a universal feature of human experience, what impact have they had on our collective history? It is to the role of these natural biases in forging the unnatural world we inhabit – from organized religion to the nation state – that we now turn.

PART TWO

Nature Extended

4.

Cranking Up Conformism

Göbekli Tepe is a series of Stonehenge-like clusters of T-shaped monoliths – each one hoisted upright like a headless giant – situated in southeastern Anatolia, Turkey. But whereas Stonehenge was built around 2,000 to 3,000 BCE, these stones were quarried much, much earlier – some 12,000 years ago. The sheer effort of human labour involved is mind-boggling. Nobody knows how it was done without bulldozers, winches, cranes, or even steel hand tools. And nobody knows why. What inspired ancient foragers to push their Stone-Age technology to the very limits, cutting twenty-ton stones out of the hillside, and carving beautiful images onto them?

When I first arrived at Göbekli Tepe, I found the monuments partially excavated on the crest of a hill, overlooking a parched and flat landscape, except at the horizon where the hills rose majestically through the haze. I tried to imagine how it would have looked 12,000 years previously. I could picture fields of wild wheat as far as the eye could see, trees with bountiful nuts and berries for the taking, and herds of wild gazelle. Right now, however, I was surrounded by scientists in broadbrimmed hats and headscarves, milling around or peering down into the excavations. Some were taking photos, others remonstrating loudly with each other about some point of scholarly disagreement, and still others clustering into small groups around one of the archaeologists working at the site, giving impromptu lectures.

I wandered through the crowd, listening in on the conversations. One of the archaeologists had picked up a piece of rock about the size of a cigar and was telling a small knot of eager listeners that it was a PPNB (Pre-Pottery Neolithic B) naviform blade, indicating that it was shaped roughly like a boat and had been made during a phase of the Neolithic before pottery had been invented. Priceless objects like this were scattered all around our feet, amid the rubble. I moved on, passing two prominent archaeologists who were arguing about whether the stele resembled the totem poles of the Pacific Northwest. And then I caught sight of Klaus Schmidt, leader of the excavations, wearing a flat cap and a light cotton scarf that flapped around him in the breeze. He was surrounded by a throng of eager listeners. I went to join them.

Schmidt was pointing to one of the great T-shaped monoliths. We were surrounded by similar monuments, each formed roughly like a giant person with arms bent at the elbows carved into its sides. Schmidt was drawing people's attention to the part of the stele where its genitals would have been, were it not for the fact that a belt-like object had been carved into it. Although we cannot see a penis, he announced, we can infer from the belt – a masculine bodily adornment – that the figure was male. As I drew closer, I could see that the monument was densely decorated with reliefs of foxes, bulls, vultures, spiders, and seemingly countless other species of fauna known to have roamed, wriggled, and flapped around the late Pleistocene or early Holocene landscape. Equally striking, however, were the images *not* inscribed on the stones. Although these complex foraging peoples were just beginning to cultivate crops, there were no plants depicted at Göbekli. And with the exception of one carving of a headless skeleton and the eerie anthropomorphism of each monolith when taken as a whole, human figures were few and far between. These master carvers were very interested in animals.

Despite controversy about Göbekli's significance, a consensus was emerging among archaeologists that this was a hugely significant ritual centre. However, why the ancient peoples needed it and what it meant remain a mystery. I was there as part of a diverse team of scientists who had been asked to help solve it. My best guess went to the very heart of the ritual mindset, a force in human affairs that can seem as vast, archaic, and submerged as the stones at Göbekli Tepe themselves. So, let's dig.

The archaeology at Göbekli Tepe represents the earliest stirrings of a revolution in the scale of human societies around 10,000–12,000 years ago. Back then, there were only about five million human beings on the entire planet – a fraction of the number of people living in many large cities today. Not only was the total population of humanity tiny compared with the billions now spread across the earth, but the groups our ancestors belonged to were minuscule compared with modern states. Until around 12,000 years ago, all humans lived by foraging for wild foods – by gathering edible plants, fruits, nuts, and seeds and scavenging or hunting wild animals. This meant being on the move, forever shifting camp in search of fresh resources. In many environments, this mode of subsistence probably sustained bands of hunter-gatherers numbering a few dozen individuals at most.

Unlike many other primates that forage in small groups, human bands of hunter-gatherers often interacted cooperatively with others, exchanging goods and perhaps joining forces in particular forms of hunting or to drive away unwanted interlopers. One theory is that these forms of cooperation would have been most prevalent among members of similar cultural groups, speaking mutually intelligible languages. But another theory is that no clear ethnic or linguistic boundaries were present among hunter-gatherer bands.[1] Of course, both theories could be correct; we should be wary of assuming a 'one-size-fits-all' view of their cultural traditions and intergroup relations. Nevertheless, we can be confident that people living in those ancient societies would have had a much more parochial view of humanity than most of us living today. For ancient hunter-gatherers, the most relevant community in day-to-day life would typically have had just a few dozen members. And even the larger network of people you may have encountered over the course of a lifetime would have consisted almost entirely of people you ran into repeatedly – allowing you to put a name to a face or to know who was related to whom.

At some locations, however, communities would experience a periodic influx of visitors – perhaps many hundreds of them – during key moments of the ritual cycle. One of those places was Göbekli Tepe. The population at Göbekli probably swelled especially when food was plentiful. It has been suggested that during the late summer and autumn when herds of gazelle passed through the region, otherwise distinct bands may have descended on this sacred site as an annual tradition – perhaps hunting cooperatively and sharing the bounty in communal feasting events.[2] It seems plausible that the ancient hunter-gatherers of western Asia were bound together into small groups, united through ties of kinship and marriage and the sacred rituals of countless local traditions. Göbekli provides a window into the flowering of this culturally rich civilization and its imagery.

But then, around 10,000 years ago, people stopped making the great monoliths at Göbekli and the site was abandoned. The farming communities that eventually spread across the lands once occupied by the creators of Göbekli's monoliths were now celebrating their newfound mastery of the natural world. They did so in an explosion of new arts and technologies, from decorative pottery to two-storey houses. Gone were the huge, occasional feasts that brought together disparate cultural groups. Instead, we find signs of a more homogenized regional culture; increasingly elaborate but lacking the reverence for nature that apparently preoccupied the hunter-gatherers at Göbekli. Eventually, everyone forgot about the monuments they had so painstakingly erected and all that remained was a sun-baked hillside in which the buried stones remained undisturbed for millennia, as if in unmarked tombs.

This chapter is about the ways in which new kinds of rituals first emerged to support cooperation in larger cultural groups – and how this gave rise to increasingly complex societies like the ones that gradually emerged after the abandonment of Göbekli Tepe. Rises in population size correlate closely with what social scientists call 'social complexity'. As settlements grow larger and denser, they need to develop increasingly sophisticated ways of managing resources, such as land to build on or farm or communal access to the water supply. Being packed into larger and more densely populated settlements requires new ways of resolving disputes, defending against external enemies, coordinating the construction of public works, and so on. The institutions that evolve to manage such problems tend to become more elaborate over time – a process that leads to more specialized roles, more widely standardized rules and norms, and eventually more centralized and hierarchical systems of governance. All these innovations are aspects of increasing social complexity.

The first step in the evolution of such complexity was the creation of larger social groups, which were organized around increasingly long-term collective planning and cooperation. And we can trace the beginning of these processes to the period following the abandonment of Göbekli Tepe. Around ten millennia ago, the world's first farming techniques were spreading – starting in the Fertile Crescent in the Middle East, and then independently emerging in regions like China and South America. Eventually, agriculture would be invented in at least eleven locations around the world at different points in time. These developments seem to have been accompanied by a fundamental transition in people's ritual lives. As groups settled and became increasingly dependent on farming, rituals became more frequent and changed form – and this in turn helped cultural groups to grow and spread.

This process was rooted in the conformism bias that I described in Chapter 1. Recall that the human tendency to overimitate is often motivated by the ritual stance: our tendency to copy the behaviours of others, even when they are assumed to lack a knowable instrumental rationale. In this chapter, we will see how that urge to conform was harnessed and extended during the transition to agriculture to build much larger-scale communities, such that the approach to ritual we have inherited today differs strikingly from that of our Palaeolithic forebears. But this is not an easy process to unravel. Understanding how ritual life was changing in the societies where crops were first cultivated and wild animals gradually domesticated presents many challenges. It requires us to look beyond the tantalizingly patchy evidence from archaeology, drawing also on what we can see in the diversity of human cultures today and taking into account what psychologists can tell us about the way humans think and behave in response to well-known environmental cues.

This is a detective story in which we must bring together all our sleuthing skills. We must consider not only the evidence to be gleaned from the crime scene itself but also our knowledge of analogous cases, probable causes, and natural biases that shape human behaviour.

Rituals, routines, and the first big groups

During the transition from foraging to farming in the region around Urfa where Göbekli Tepe is located, archaeologists are unsure about the nature of the groups people belonged to. When groups coalesced at sacred sites, did they form a single community or were these gatherings composed of rivalrous tribes, with distinct group alignments? It is possible, for example, that each of the mighty stone slabs erected at Göbekli served as a totem for a different group, each associated with its own god or ancestral founder, and that the very act of creating monoliths could have been conducted competitively – each tribe seeking to outdo the others in displays of strength and cooperative capacity as an expression of local pride and dominance or prestige in intergroup rivalry.

But whatever the details of these cultural groups, they were likely small-scale and localized. Although monoliths similar to those at Göbekli Tepe could be found not far away – suggesting that their creators participated in a common cultural system – anyone travelling further afield would have found stele that looked quite different, as well as regions that had none at all. Not only were the geographical bounds of the culture quite limited but they also faded over time. Archaeologists once thought that the cultural groups responsible for the T-shaped monoliths died out, only to be replaced much later by cultural systems spreading into the region from elsewhere. But recent discoveries suggest a different story.[3] Findings from Tell

Halula in northern Syria point to continuities between the peoples of early Neolithic Urfa, who created the stele at Göbekli, and the cultural systems that are visible in the archaeological record much later. This suggests that the cultural systems that eventually prevailed in the region surrounding Göbekli were not suddenly transported there from afar but evolved slowly through the efforts of Göbekli's descendants.[4]

Although the precise details of this process remain obscure, it is clear that groups sharing a common culture became very much larger over time.[5] The distinctive pottery designs and stamp seals of the Halaf period encompass not only the region of Turkey in which the cultural system at Göbekli once flourished but huge regions of what are now Syria and Iraq as well. What changed? How did very localized cultural groups evolve into much larger ones sharing the same norms, rituals, customs, hairstyles, stories, and culinary traditions?

My own answer to this question, once again, has its origins in my time in East New Britain, Papua New Guinea.[6] Prior to the arrival of colonial administrations and missionaries in the nineteenth century, the peoples of East New Britain were living in cultural groups that were no larger than those of Urfa's Neolithic prehistory and perhaps much smaller. Many of the languages of the region were mutually unintelligible and relations between the speakers of those different tongues were often precarious and potentially violent. Elders in the village where I lived told me that encounters commonly led to ritualistic beheadings and cannibalism. In some cases, the more highly organized groups took captives from smaller-scale groups as slaves or unwilling brides. Even among those who spoke the same language, albeit perhaps adopting distinct dialects, warfare and raiding was at least as likely to occur as peaceful trading or intermarriage.

I was particularly interested in the form that ritual took in these societies. As we saw in the last chapter, I had become preoccupied with the way the warriors in these groups were bound together by painful and gruelling ritual ordeals. Crucially, though, these rituals took place quite irregularly. Even the most frequently occurring collective rituals happened only occasionally – for example, to celebrate a major festival or communal achievement. These relatively rare events were major highlights in the ritual calendar, often accompanied by feasting, dancing, and singing. Was this also the way in which sporadic rituals functioned at ancient sites like Göbekli? And if so, what impact would this have had on the scale and structure of cultural groups?

I argued in Chapter 3 that infrequently performed rituals forge especially strong bonds between their participants. These occasional, emotionally intense moments become defining features in our memories, helping to fuse personal and group identities together. The groups formed in this way tend to be relatively small-scale and localized. Even if to a casual observer they have some cultural practices in common with neighbouring communities – such as food preferences or architectural styles – each group's most potent symbols of identity are strikingly distinct. For example, although the great monoliths of Göbekli are shaped like a letter 'T', settlements a little further away at Çayönü, Qermez Dere, and Gusir Höyük had stone monoliths but without that distinctive shape. It is possible that people living quite close by had very different cultural traditions – perhaps knowing about each other's monoliths but nevertheless insisting on doing things differently.

Likewise, many of the distinct groups living in New Britain differentiated themselves from their neighbours by performing highly idiosyncratic versions of otherwise quite similar types of rituals. For example, many Baining communities performed distinctive fire dances, where initiated warriors wore highly elaborate masks and hurled themselves into the flames while large groups of men beat out rhythms with lengths of bamboo and sang together at the tops of their voices. But each group had distinctive masks and its own rhythms and songs, distinguishing itself from all others. And the secret stories, myths, and symbols that lay behind these traditions were all highly localized – specific to the groups that sustained them.[7]

The ancient foraging peoples of Urfa, just like more recent

generations of indigenous tribes in East New Britain, would no doubt have taken pride in the specialness of their own local communities, celebrating their differences and perhaps downplaying similarities. In New Guinea, the roots of these cultural differences originated in small innovations as groups split off and spread across the landscape. Minor departures from the script often went unnoticed. But the further you got from the place where the tradition began, the more your version of it would become a copy of a copy of a copy – each deviating slightly further from the original. Every performance of a ritual and every telling of a myth carries the risk of garbling and forgetting.[8] As each group tries out slightly new versions of its existing stock of rituals, it introduces small deviations which, over time, eventually lead to a great diversity of ritual groups across the landscape: a kind of patchwork quilt of local traditions, like those to be found in regions like Papua New Guinea.[9]

All this changed dramatically with the arrival of Catholic missions (among other Christian denominations) to East New Britain, which insisted on much more regular forms of worship. The Roman Catholic Church rapidly established standardized versions of the Eucharist, baptism, communion, confession, prayers, hymns, doctrines, and narratives, that were nearly identical not only in New Guinea villages but across a dizzying variety of countries spread around the globe. They were carried out in similar-looking social settings in Latin America, Ireland, eastern Europe, India, China and who knows where else, week in and week out.

How did Catholicism manage to turn the manifold rituals of East New Britain into a single, standardized form of worship – mirroring practices being carried out in churches thousands of miles away? Clearly, part of the explanation for this standardization is that the Roman Catholic tradition is highly organized, with an elaborate ecclesiastic hierarchy regulated centrally from the Vatican. Another part of it is that the beliefs and practices that devout church-going Catholics should carry out have been inscribed in texts – many of which have venerable histories and are regarded as sacrosanct. However, there is also a less obvious reason why such religions have become standardized across huge populations scattered over vast territories. That reason is *routinization*.

Routinization refers to the continual repetition of beliefs and practices on a daily or weekly basis. In pre-missionary New Guinea, rituals were infrequent and emotionally intense. With the arrival of Catholicism, however, some key elements of ritual life became more frequent and less intense – they became routinized.

On the face of it, routinization may seem very puzzling. Have you ever wondered why the most globally widespread religions require their adherents to attend temples, synagogues, mosques, churches, and other sacred buildings to carry out precisely the same procedures day after day, week after week, month after month, year after year? This is a peculiar state of affairs. If you are seated in a pew in any Catholic church on any given Sunday morning anywhere in the world you will find yourself chanting phrases and repeating actions that you have previously performed on so many occasions in the past that you know them all by heart. You are not learning anything new by doing these things, and it is hard to fathom why the God to whom these behaviours are directed should need things to be repeated so many times. Surely God – especially one considered to be omniscient – would not need to hear the same old prayers and hymns over and over? From a practical perspective, the behaviour seems quite redundant both from the perspective of the actor and the intended recipient. Worse still, such frequent repetition might reasonably be said to impoverish any sense of spontaneity or sincerity in the acts of worship. That is, the extraordinary familiarity of the routines makes it somewhat inevitable that many worshippers will carry them out on 'autopilot' without any need for conscious reflection or engagement.

But in fact this kind of extreme repetition is anything but redundant. While routinized rituals might seem like a waste of time and resources, they are actually one of the most extraordinarily potent forms of social control ever invented by human groups.

Routinization does several consequential things. Firstly, it turns us into extreme conformists. In Chapter 1, I explained that we copy people more fastidiously when we think that the behaviour in question has no instrumental function whatsoever: when we adopt what I call the 'ritual stance'. It is hard to imagine a more potent signal of this than the requirement that you repeat the same actions and chant the same words over and over, way beyond the point of redundancy. So, one effect of routinization is that it promotes more faithful observance of tradition. Secondly, it makes us less reflective and therefore more susceptible to authoritative pronouncements. The more frequently we conform to a ritual script, the more automatic and habitual the behaviour becomes. And the less we reflect on *how* to carry out the procedures in question, the less we think about *why* we are doing them at all. This makes us more susceptible to being told what they mean and accepting this without thinking critically for ourselves. This too is an important foundation for establishing very high levels of standardization in a belief system and low levels of innovation and independent thinking.[10]

All this may sound like a bad thing. In a culture that values individualism, creativity, and thinking for ourselves, routinized ritual and conformism may seem highly undesirable. But this formula has proven to be remarkably useful to human groups over the course of world history. Indeed, I would argue that the appearance and spread of more routinized rituals opened the door to cooperation in much bigger societies. Why? Because it makes it possible to stabilize cultural traditions across larger populations and territories.

I first began to understand how this works in practice when I considered the impact of routinization on the religious culture of the Baining and their neighbours. The power of routinization to unify rural peoples came as a huge revelation to followers of the Kivung, the new religious movement I was spending most of my time studying. For anyone joining the Kivung for the first time, one of the most striking consequences was that it required participation in an extremely frequent round of rituals. Indeed, it was so time-consuming and labour-intensive that people frequently talked about the 'work' of the Kivung as if it were equivalent to a full-time job. For many followers of the movement, that was no exaggeration.

Laying out offerings, performing rituals to achieve catharsis and absolution, cleaning the graves of the dead, meeting to discuss issues of morality, theology, and cosmology – all these activities were organized around cycles of daily, twice weekly, weekly, fortnightly, and monthly repetition according to a strictly enforced schedule. To ensure that everyone knew when it was time to perform another crucial ritual, a gong would be struck repeatedly through each day to summon people from their gardens, cash crops, and domestic duties. The 'gong' might be any kind of object capable of creating a loud noise – from a traditional slit drum to a discarded metal cooking pot, if available. Where I happened to live, the village gong was the shell of a large bomb from the Second World War that had been found in the forest and was now suspended from a tree in the centre of the village.

While I was collecting life histories from Kivung followers, they would often tell me how impactful the new routinized lifestyle had been for them and how it had expanded their social universe. When people first joined the Kivung, they were initially daunted by the demands it would make on their time. Many were unwilling to make that sacrifice and decided not to take part. But often villages joined up as entire communities and it was harder to avoid following the herd. Having joined, however, people realized that they could now travel to scores of other villages across the region they had never visited before and find themselves welcomed there, treated hospitably, and invited to participate in the same rituals that had become familiar at home. This made it possible to develop greatly expanded networks of acquaintances in ways that brought social, material, and other benefits, even in communities that spoke different languages and would traditionally have been regarded as enemies. Thus, the Kivung unified thousands of people in an unprecedented way. And it was able to do so precisely because of its high-frequency rituals. Never before in this region of Papua New Guinea had it been possible to create such a common system of beliefs and practices across such a wide region and to unite it under the banner of a single identity.

This system supported the emergence of the first 'imagined communities' – groups too large for their individual members to know each other personally. Think of your country, for example. That is a meaningful group category for most people living today, but you couldn't possibly know all its members and so in a sense it is a community of imagined others, rather than of flesh-and-blood persons you could point to more concretely.[11]

The introduction of routinized rituals to Kivung villages in the 1960s had an almost instantaneous effect on the unification of various previously fragmented villages and language groups in the region. In time, this newfound shared identity also became a powerful force in elections, enabling Kivung-backed politicians to gain seats in both national and regional assemblies. And these new forms of political alignment persisted, remaining strong in the 1980s when I carried out fieldwork there. But these time periods are minuscule compared with the many thousands of years it took for our ancient ancestors to discover the unifying power of high-frequency collective rituals. For most of human prehistory, unification via routinization was an unimaginable idea – in much the same way as writing, which would arrive later still. Both would eventually allow human culture to be stored and spread in a more stable form than ever before. This process was only able to unfold so quickly in Papua New Guinea because Christianity and its techniques of schooling and regular worship provided a ready-made blueprint for the creation of routinized rituals.

My experience in East New Britain made me wonder whether the routinization of ritual had played a similar role in ancient societies. Was the invention of routinized ritual a key force in the emergence of the first large-scale communities? As it turned out, this was not a straightforward question to answer. We know how routinized rituals and their consequences were introduced to the peoples of Papua New Guinea, thanks to extensive documentation of the efforts of Christian missions.[12] A far greater challenge was to figure out how routinization was invented in the mists of prehistory and how its consequences were discovered and exploited by ancient societies, long before their beliefs and practices could be written down. If routinization was indeed a driving force in the emergence of the first large-scale societies, how might we tell?

The birth of routinization

In principle, it makes perfect sense that the gradual appearance and spread of agriculture – the so-called 'Neolithic revolution' – would have given rise to more routinized collective rituals. Farming meant that various domains of human activity became more elaborate. Suddenly, you weren't just killing animals or gathering nuts and berries. There were now myriad stages in the production process: from the extraction of raw materials to the fabrication of tools, from the planting of seeds to the harvesting of crops, from the winnowing and grinding of wheat and barley to the cooking of ever more elaborate meals. Many of these tasks required frequent repetition, as part of an unremitting daily round of workaday activities. In turn, these routinized skills – many of which became embellished with conventional frills such as the frequent replastering of walls and the creation of clean spaces in the houses where food-processing and cooking took place – would naturally come to form a unique 'way of doing things': a routinized cultural system embedded and enacted in everyday life.

But this routinization of ritual and custom would not simply have been a useless *by-product* of the transition to agriculture. It would also have brought many benefits associated with the standardization of a shared culture. As people increasingly crowded together in larger and denser settlements, all depending on the same finite resources, social relations would likely have come under increasing strain. Relative strangers would need to be able to interact cooperatively to an extent that was previously unimaginable. Having a common culture and identity would have made this much easier to accomplish. Recognizing that relative strangers are like oneself would have made it easier to communicate courteously instead of butting heads.

As such, farming and routinization were likely to have been mutually reinforcing as agriculture became established. Just as agriculture created more repetitive daily routines – many of them becoming ever more ritualized – so the farming lifestyle was supported by larger-scale forms of group identity and cooperation.

Or that was the theory, anyway. But how could we possibly hope to see any of this in the archaeology of ancient societies, long before writing was invented? The situation is not as hopeless as it may seem. Just over 400 miles to the west of Göbekli, in southern Anatolia, there is another important archaeological site known as Çatalhöyük which became established centuries later. There, at Çatalhöyük, it is possible to observe changes in the ritual lives of the inhabitants and in the scale on which group identity was being forged.

The site comprises twin mounds of earth and rubble created by human activities spread over around 2,000 years, starting more than 9,000 years ago. The reason the mounds came into being is that each new house was built on the foundations of an older house that had stood there previously. After many generations, this resulted in houses being constructed at ever-higher elevations, creating hills. But it also meant that drilling down through the rubble was like travelling back in time, through the generations of people who lived there.

In an effort to trace the first emergence of routinized rituals in human history, I became deeply preoccupied with Çatalhöyük and ended up visiting the site many times – during one phase travelling there six summers in succession. Summer was always the digging season, and I spent many scorching weeks following the archaeologists around, badgering them with incessant questions. My hunch, which grew stronger the more I learned, was that this site contained the key to understanding the origins of large-group living. It seemed my theory was onto something: just as frequent rituals in the Kivung had allowed the Baining to create a new identity, encompassing populations previously divided into distinct and competing groups, something quite similar seemed to have been happening thousands of years ago at Çatalhöyük, albeit much more slowly. In my efforts to understand this process, I began working with the director of the site, Ian Hodder.[13]

One of the biggest challenges faced by archaeologists working at Çatalhöyük has been to establish patterns of continuity and change over the period of settlement.[14] For example, were people's diets changing and, if so, was this related to changes in patterns of hunting, gathering, crop cultivation, and animal domestication? What was happening to population size and density? Was working life much the same throughout the millennia or was it getting easier or harder, simpler or more specialized? And, of particular interest in my collaborations with Hodder was the question: do we see evidence of changes in ritual life – and, in particular, were collective rituals becoming more frequent and standardized?

Answering these kinds of questions based on the interpretation of objects that have been buried in the ground for many thousands of years is not easy. But the picture emerging at Çatalhöyük is becoming clearer as more and more of the archaeology is brought out into the light. Over time, numbers of domesticated sheep, goats, and cattle increased and reliance on hunting declined. The volume and variety of artefacts found in houses also increased but so too did the amount of labour required to make and maintain those artefacts. This indicates that even though the material culture was expanding, people were also working harder. Daily life seems to have become more burdensome and analysis of human remains indicates that there were more accidents and injuries associated with the increased workload.

So, how was all this related to the ritual lives of these ancient people? Some of the most interesting evidence on this topic has emerged from going through people's rubbish. The inedible remains of animals eaten at Çatalhöyük were discarded after all the nutritious material had been extracted and the resulting midden deposits provide quite detailed information about what kinds of meals people were eating, and when. As a result, we can see that the rituals surrounding people's eating habits were changing dramatically through time.

Consider their approach to feasting. It is clear that during the earliest phases of settlement, the meat of wild bulls was consumed in large feasts and the horns of these beasts were frequently embedded in the walls – like the trophy heads adorning the hallways of English country mansions. Except that the bucrania on display in the houses of Çatalhöyük were not shot at a distance by men in tweeds and plus-fours but killed in close proximity at risk to life and limb, requiring considerable daring and courage. Wild bulls in Neolithic Turkey were huge and aggressive animals, measuring roughly six feet at the haunches – far larger than their domesticated descendants. And yet, if the images of hunting depicted in the wall paintings at Çatalhöyük are to be believed, they were brought down by teams of thirty or more men attacking from all sides with sharpened projectiles, then attempting to ensnare and bind the animal's feet with their bare hands. We cannot know how many people were injured or killed in such endeavours, but we can speculate based on the wall art that their bravery was celebrated in massive community feasts and ritual memorials. We can also infer from the remains left over from feasting events that these great communal ceremonies were only sporadically enacted – at most, once in a decade or so and perhaps only experienced by many once in a lifetime. In these and other respects they were typical of the 'imagistic' rituals we have met already: emotionally intense but only rarely performed.

Hodder and I concluded that imagistic rituals reached their high water mark around the middle of the period of settlement at Çatalhöyük, at which point we find evidence for the most lavish feasting events and elaborate displays of bucrania in houses. In many ways this resembles the patterns observed at Göbekli Tepe – except that instead of the community struggling to erect huge monoliths and adorning them with images of wild animals, the people at Çatalhöyük were wrestling to the ground actual flesh and blood animals and dragging them back to the feasting ground, memorializing the event not in stone but in the trophy heads of animals installed into their walls. It seems plausible that in both cases, the communities bound together by these activities were limited to the relatively small number of those who participated. These sporadic, emotionally intense rituals were leading to small, tightly bonded communities.

But meanwhile in the background a massively consequential change was taking place at Çatalhöyük – one that does not seem to have happened at the earlier site of Göbekli Tepe. Rituals were becoming more frequent. The very earliest evidence of ritual routinization at Çatalhöyük was only modest and easily overlooked. It took the form of a growth in all kinds of everyday tasks. Although many aspects of those tasks were surely instrumental – for example, designed to improve the efficiency of subsistence tasks such as grinding flour, butchering animals, weaving baskets, or cleaning the hearth – they also likely incorporated an ever-increasing diversity of causally opaque conventions: in short, integrating rituals into the most everyday of objects and activities. Now, baskets were not just mundane receptacles for gathered foodstuffs – they were also decorative objects coding socially important information about their makers and owners. Cleaning the house, especially certain spaces within it, became almost an obsession – way beyond what might reasonably be required for any practical purpose. For example, houses contained raised platforms typically in the north or east parts of the floorplan that were kept scrupulously free of debris (such as food remains, ash from the fire, bits of broken tools, or other evidence of domestic work). And the inhabitants of these dwellings seemed preoccupied with continual replastering of walls in a white marl slurry, applied with great care and attention perhaps as frequently as every month. Ritual had entered people's houses, and as rituals in these spaces became ever more prominent, family homes became increasingly like miniature temples and places of worship.[15]

This process of growing ritualization at Çatalhöyük evolved over scores of generations. And in the process, the nature of these routinized rituals changed. By the later stages of settlement around 7,000 years ago, the volume and variety of domestic tasks, both ritualized and instrumental, had begun to move from internal to external spaces. People's routines became increasingly visible to the wider community. In tandem with this shift, we see the increasing standardization and stabilization of various everyday artefacts – from stamp seals to pottery designs.

Hodder and I believe that this shift towards a more homogeneous culture across the settlement as a whole was linked to a transformation of ritual. By now, sporadic, emotionally intense rituals were no longer the main form of social glue holding the group together; in their stead came more frequent, mundane domestic rituals. Feasting events became rarer as people became increasingly reliant on domesticated animals as their main source of meat. The heroic events of yesteryear became the stuff of collective storytelling, rather than lived experience. Instead of teasing and baiting powerful wild animals in hair-raising collective rituals, it seems people were now increasingly just telling stories about past events, turning them into myths and pointing to the heads of long-since slaughtered beasts embedded in the walls of their houses.

The people of Çatalhöyük could not have known how consequential this shift of gear in their ritual lives would prove to be, but I believe it constituted a quiet revolution in the prehistory of humankind. Their transition to an increasingly routinized ritual life would have changed the way people saw themselves as members of a community, paving the way for the emergence of much larger group identities.

Routinization and the rise of social complexity

The settlement at Çatalhöyük is just one, early example of how the transition to agriculture gradually brought new forms of ritual, and in turn larger cultural groups. Over time, these same processes would intensify, leading to much larger-scale civilizations around the world.

For example, one of the first large-scale, complex societies in world history, capable of unifying cities located hundreds of miles apart, was the so-called Harappan civilization which spread along the Indus Valley (in what is now Pakistan) more than 4,500 years ago. One of the hallmarks of this society was an extraordinarily high degree of standardization, apparent in its iconography and architecture and other indicators of shared beliefs and practices, especially ones associated with bodily purity and public sanitation. Harappan seals – used to mark pots in a highly consistent way across the civilization – provide a good example. Seals were designed in a virtually identical fashion across the entire empire, depicting a line of (as yet undeciphered) script along the top, a species of animal in the middle, and additional inscriptions along the bottom. Likewise, buildings were constructed in the same way across all Harappan cities and flood defences requiring labour forces of thousands constructed according to exactly the same principles of building control.

You might imagine that such a degree of cultural homogeneity would have required some kind of hierarchical system of governance. Unlike many later empires, however, this early civilization appeared to lack a system of top-down administration and control. Harappan methods of town planning and water management would have required highly sophisticated forms of cooperation and coordination, but this did not seem to entail a system of social stratification with powerful leaders at the top and slaves at the bottom. Nor was conformism enforced through military domination. The Harappans appeared to lack armies or elaborate weaponry, being more interested in trading with surrounding groups and procuring luxury goods from other centres, such as in Mesopotamia, by peaceful means. In short, the force binding together Harappan culture seems to have been neither violence nor centrally organized bureaucracy – but routinized ritual. The high levels of standardization observed in this civilization suggest that its norms, beliefs, and practices were embedded in the repetitive practices of everyday life, particularly ones associated with cleansing and purification. Much like the people of Çatalhöyük before them, and East New Britain long after, routinization arguably helped create a new cultural group, one larger than any that had existed in the region before.

As rituals became more routinized and societies grew larger, more intensive farming would have been needed to feed more densely settled populations. Some of the best evidence for this comes not just from archaeology but from the ethnographic record. When I moved from Belfast to Oxford in 2006, I brought with me a grant that was focused on explaining various patterns of recurrence and variation in the religious beliefs and practices of human societies in modern times. Because of my interest in the unifying effects of ritual routinization, a high priority of this project was to explore where and why high-frequency rituals appear in cultural groups around the globe. Using this grant, I was able to hire a postdoctoral researcher, Quentin Atkinson, to work with me to build a database of rituals from a representative sample of the world's cultures and to measure their frequency of performance and emotional intensity. Atkinson had mastered the statistical techniques necessary to establish whether ritual frequency was indeed related to group size: with emotionally intense, infrequent rituals – like the hunting feasting events at Çatalhöyük – being associated with small groups; and emotionally less intense day-to-day rituals – like the increasing routinization of domestic rituals over time – being associated with larger groups, as I had hypothesized.

To explore these ideas, Atkinson and I first set out to test whether infrequent 'imagistic' rituals tend to be more emotionally intense, and frequent 'routinized' rituals less so. To this end, we created a database encoding information on about 645 rituals from a sample of seventy-four contemporary cultures from around the world.[16] We sourced this information from a massive repository of writings by ethnographers known as the Human Relations Area Files,[17] together with an additional information source known as the *Ethnographic Atlas*,[18] which recorded much relevant information on an overlapping set of cultures pertaining to variables of interest such as group size and structure. From these sources, we extracted the information we needed on the frequency of ritual performances, their emotional intensity, the size of the groups they occurred in, as well as various other features that we thought might be interesting to explore, including various aspects of economic life and subsistence strategy. When Atkinson completed his statistical analyses of the data, it revealed that, as predicted, rarely performed rituals tended to involve high levels of emotional intensity. But something even more exciting came out of the analysis of our rituals database: the discovery that *routinization was associated with agricultural intensity*. In other words, as societies become more dependent on farming, their rituals become more frequent.

In turn, this finding triggered an even more challenging idea. Would we find similarly convincing quantitative evidence that the performance of more frequent rituals was enabling larger-scale cultural groups among early farmers to stabilize and spread through history? This thorny question led archaeologist Mick Gantley to come up with a novel methodology that we termed 'Material Correlates Analysis' – or 'MCA' for short.[19] The MCA approach was inspired by the rituals database that Atkinson and I had created using ethnographic writings as our source of information. That database had found a strong link between routinization and agriculture in contemporary cultures; but Gantley also set himself the task of finding out whether we could use this information to shed light on how ritual life was changing in our ancient past, during the agricultural transition in western Asia.

The central idea of the MCA approach is that there are numerous features of material life that are visible in the remains dug up at hundreds of ancient archaeological sites in the region of the world we wanted to survey. So, even though there are many things we cannot see directly, we can at least compare well-documented features with those of societies present in our sample of more-or-less contemporary societies in the rituals database. Doing so produced ninety features that could be compared directly from the observable archaeological remains and from a sample of extant ethnographic cultures. Such features included details of how food was produced and consumed, how communities were organized, and what sorts of rituals people were performing. These constituted our 'material correlates'. Based on this methodology it became possible to say which ancient societies more closely resembled which modern ones. It was only a short step from doing that to asking the question: which ancient societies are more like contemporary 'imagistic' ones – such as the cults of New Guinea tribes involving rare but emotionally intense initiations – and which societies more closely resemble the modern ones with more routinized rituals – such as today's world religions which emphasize regular worship in special temples, churches, mosques, and synagogues?

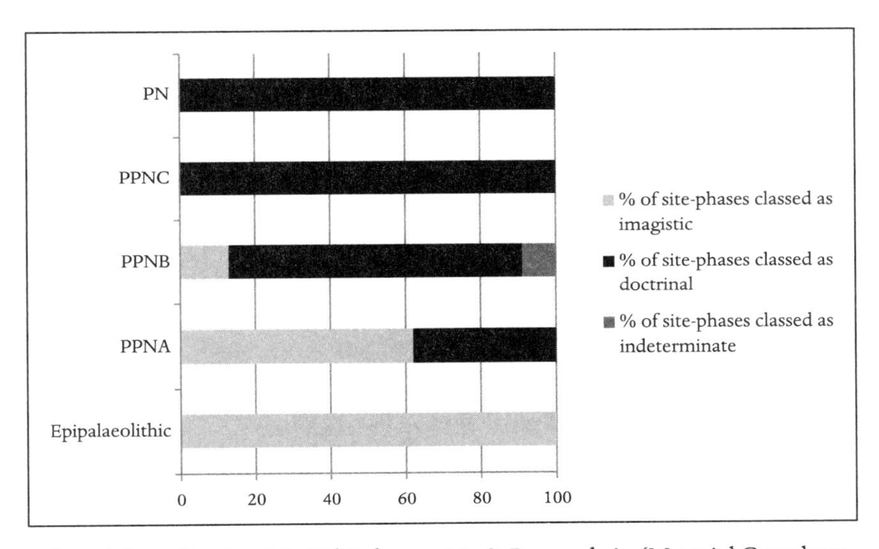

Adapted from Gantley, M., Whitehouse, H., & Bogaard, A., 'Material Correlates Analysis (MCA): An Innovative Way of Examining Questions in Archaeology Using Ethnographic Data', *Advances in Archaeological Practice*, Vol. 6, No. 4, pp. 328–41 (2018)

The findings were striking. This diagram shows the results of our efforts to compare a sample of forty-nine archaeological sites from western Asia spanning the agricultural transition.[20] Archaeologists generally show more ancient phases at the bottom of their diagrams (corresponding to the lower levels at which older objects are unearthed) and more recent phases at the top (where objects in the ground are generally found nearer to the surface). So, the most ancient level depicted here is the 'Epipalaeolithic' (the very final throes of the Palaeolithic era before the invention of farming, when all humans subsisted on wild foods) and in between are various periods described as 'PPN' (Pre-Pottery Neolithic, divided into phases A, B, and C) or 'PN' (Pottery Neolithic – when pots had finally been invented). During the Epipalaeolithic period, the light grey colouration indicates that, according to Gantley's statistical analyses, ancient foraging cultures were predominantly imagistic – they were bound together by rare, emotionally intense rituals. By contrast, the most recent periods of the Neolithic at the top of the diagram marked in dark grey, when agriculture was well established, had the most in common with modern societies – they adopted more routinized rituals, with all their potential to bind together larger groups of people. During the transition from imagistic to doctrinal modes we found some more indeterminate practices (medium grey). This provided quantitative evidence that ritual was becoming more routinized during the agricultural transition – not only at Çatalhöyük but across a much wider region.

Farming and future-mindedness

I have argued in this chapter that more frequent collective rituals contributed to the rise of large-scale groups during the Neolithic revolution by helping to stabilize cultural identity markers: or, to put it more simply, that routinized rituals helped create bigger groups. But there is another, as yet unexplored reason why routinized rituals would have been useful to early farmers: and that is how they influenced our attitude to the future.

Farming and future-mindedness go hand in hand. I first learned this during my first week at university as an eager eighteen-year-old undergraduate. One of my teachers – James Woodburn – had shocked the anthropological community by returning from his fieldwork among the Hadza of Tanzania, declaring that he had found a cultural group with no social organization. His colleagues at the London School of Economics still talked about the hubbub this caused. The cries of disbelief and the murmurings of scepticism continued to echo down the corridors of the anthropology department.

When asked about this matter directly, however, Woodburn said his colleagues were exaggerating. It was not exactly that the Hadza lacked social organization, but that their social groupings were highly flexible – people could come and go as they pleased and there were no strong expectations of group loyalty or reciprocity. Food tended to be hunted or gathered informally on a day-to-day basis and was shared with little concern for who received what – a form of pooling that Woodburn likened to a system of taxation in which the unproductive were subsidized without sanction or stigma. The very young or elderly tended to benefit, but this evened out in the long run because they would either contribute more in the future or had done so in the past.

This helped to make Hadza society very equal. All the kinds of resources that would normally be used as a basis for social inequality – wealth, power, and status – were quite evenly distributed. Food and possessions could not be stockpiled because the Hadza were nomadic, and they needed to travel light. Nobody could boss anybody else around because everyone had access to the same basic weapons. Gender inequality was also minimal because women tended to form solidary support groups at the level of the band rather than being separated off into domestic units, potentially dominated by patriarchal or aggressive husbands. And differences of status were actively prevented through the application of what Woodburn referred to as 'levelling mechanisms' – norms that actively inhibited expressions of pride and competitiveness. Woodburn described these societies as 'egalitarian' and he referred to their mode of subsistence as an 'immediate-return' economy.

Woodburn had highlighted one of the key distinctions between simple hunter-gatherer societies and more complex ones. If you make your living from hunting and foraging in small bands, the prospect of a bigger reward in the future is usually less compelling than the assurance of a smaller reward right now. Woodburn observed that more complex hunter-gather societies – such as those found on the northwest coast of Canada – had a delayed-return economy depending on elaborate dams and weirs which required careful forward planning and coordinated effort. But what struck me as I listened to Woodburn's arguments was that the shift away from short-term thinking was far more pronounced and enduring in societies that abandoned hunting and gathering altogether and took up farming. Farmers must wait patiently for crops to grow and for domesticated animals to mature before they are eaten and so a hand-to-mouth existence is no longer viable. A more future-minded approach to life is required – not only in order to ensure that the food supply is available when needed but also, as we shall see, to facilitate its distribution in a way that ensures everyone gets to eat.

In this way, the invention of farming led to a newfound dependence on technologies of food production and storage that made planning ahead absolutely essential. Once you have to wait for crops to grow and ripen – and once ownership of produce is associated with individual families rather than the entire tribe – it becomes very difficult to sustain an 'immediate return' lifestyle. It is now necessary to store food for times of hardship or to find ways of ensuring that surpluses are continuously being moved around, so that the better-off can subsidize the less well-off in times of need.

Delayed-return economies introduced new challenges for human societies. One of the most basic was the need to ensure that the whole community would be sufficiently well provisioned that nobody would go hungry during lean periods. Archaeologists trying to understand how this problem was addressed in Neolithic prehistory often refer to a book entitled *Stone Age Economics* – written not by a specialist in some field of prehistory but by an anthropologist like me, called Marshall Sahlins.[21] One of Sahlins' great insights was that the rise of agriculture had a significant effect on the social units that made up a community. He argued that agriculture favours a 'domestic mode of production' in which the household becomes the primary unit of production, consumption, and exchange. Instead of the band functioning as the main economic unit for procuring and sharing the resources needed for survival, families became settled into distinct dwellings, formed around a common hearth rather than a collective campfire.

But while this is an effective way of organizing small-scale farming, marriage, and child-rearing, it also has certain downsides. Perhaps the biggest problem with the domestic mode of production is that all households are at some point unfavourably composed in terms of the ratio of effective producers to dependent consumers. Young households have many hungry mouths to feed and relatively few capable adults. Long-established households have doddery founders whose more capable dependants have flown the nest and are taken up with the challenges of supporting families of their own. If domestic units were left to fend for themselves in such a system – based on immediate-return thinking – it would soon lead to disaster in the form of widespread household failure. Such communities needed to develop ways of organizing households into larger support networks: systems capable of encouraging stronger households to subsidize the ones that might be ailing.

There is some debate among archaeologists about exactly when the household first emerged as the main unit of economic life in the Neolithic.[22] Some suggest that the evidence for autonomous households can be traced back to very early in the Neolithic around 10,000 years ago,[23] while others argue it was closer to 9,000 years ago – around the time of the invention of pottery,[24] and this also seems to be consistent with the evidence from Çatalhöyük.[25] But whichever account we follow, it seems that the household came to be an increasingly important feature of economic life among settled farmers, and this meant that the risk of household failure needed to be addressed. Somehow, these ancient Neolithic societies needed to find a way of consistently thinking ahead like never before.

In the case of our Neolithic farmers, there seems to have been a cluster of innovations that helped address this problem. One of these was the rise of the so-called 'history houses'.[26] This term alludes to the way households became the repositories of especially rich and varied material culture thought to be associated with the telling of stories about the past. The distinguishing feature of these houses is that they contained more human remains buried under the floor, more artworks painted on their walls, or more heads and horns of animals embedded into the plasterwork. For example, at Çatalhöyük I saw evidence that human remains had been extracted from under the floors of houses where they had been buried, not only once but multiple times and not only within recent memory of the primary burial but sometimes many years later, after most or all of those who would have known the deceased had themselves passed away. It is easy to see how this would have promoted a more future-oriented outlook: by creating more elaborate narratives about the past, people came to see themselves as part of a long-term and unfolding history, which existed before they were born and would continue after they died.

The impact of historical narratives on legacy thinking is difficult to demonstrate using direct evidence from archaeology alone. However, another plausible indicator is the growth of high-frequency domestic rituals at Çatalhöyük. As we will see in Chapter 7, we now have experimental evidence that participation in routinized rituals has a remarkable knack for helping people defer gratification. In other words, routinized rituals make us more future-minded. This suggests that the increasing frequency and standardization of ritual life over the course of the Neolithic would also have fostered growing concerns with future provisioning and planning.

But if history houses and routinized rituals encouraged people to look ahead more than in the past – and therefore to prioritize future provisioning more than their foraging forebears had done – they did not provide a mechanism to actually share agricultural surpluses with vulnerable households. Concretely, what was to be done about those elderly or childbearing families who couldn't produce enough food to get by? This is where another key innovation came in: a greater focus on collective rituals and community-wide obligations.

Archaeologists have argued that any tendencies towards the atomization of households in the Neolithic would have been mitigated by competing inducements to contribute to community life. For example, upon visiting another Neolithic site not far from Çatalhöyük, I discovered that even though the household emerged as a key economic unit, there was also strong evidence for activities operating at a higher level. There I found evidence for large-scale collective activities: outdoor ceremonies, the sharing of meat, cooperative foraging and farming, and collective rituals, dances, and feasting events.[27] Through these rituals, the tendency towards the isolation of individual households would have been counteracted by obligations to circulate surpluses around the community.

In some cases, these activities seem to have been rooted in elaborate gift-giving rituals, in which different households felt socially compelled to offer their surpluses to others as gifts. Indeed, Sahlins suggested that the main way in which ancient societies solved the problem of household autonomy and the domestic mode of production was through the rise of superordinate forms of cooperation, functioning above the level of individual domestic units – from household clusters[28] to much larger descent groups[29] – and chief among these forms of cooperation was an emphasis on reciprocal gift-giving.[30] Although evidence for gift exchange in Neolithic societies is limited in the surviving archaeology, it seems likely that reciprocal obligations between households were among the factors driving surplus production and the circulation of goods.

We now operate in a world in which the need for future-mindedness and community-level organization is so self-evident that it is difficult to comprehend the significance of these changes during the Neolithic. But between them, these convergent innovations – history houses, more routinized rituals, and community-wide obligations of reciprocity – seem to have helped overcome the biggest problem facing the earliest farming communities: the fact that a purely household-based economy could not survive without some higher level of cooperation and forward planning. History houses fostered future-mindedness while community traditions and obligations created a mechanism to share food surpluses with families who might otherwise go hungry. Routinized rituals meanwhile not only helped the first agricultural communities grow in size; they helped them think ahead more effectively than ever before.

The positive role of rituals in human history has taken me a long time to appreciate, in part because I have had a difficult relationship with rituals personally. This is not so unusual – many of my fellow academics study the things they find particularly difficult to master or understand. Political scientists often strike me as poor strategists in departmental politics, anthropologists struggle with the nuances of their own native groups, and geographers are often the colleagues most in need of directions around campus. Although my job is to study rituals, I find them rather aversive – particularly the routinized kind we were all obliged to participate in at school in the form of assemblies and chapel services. These activities seemed at the time to be not only pointless and dull but sometimes even menacingly oppressive and authoritarian. I am not alone in thinking this way, of course. As we will see in the final part of this book, many societies are shedding their ritual traditions at a rapid pace, not only through secularization but by dismissing as irrelevant or oppressive many long-established rituals associated with schooling, professional guilds, the institutions of government, and everyday domestic life.

However, the more I have learned about the effects of participation in frequent collective rituals, the more I have come to appreciate their importance in fostering large-scale cohesion, cooperation, and future-mindedness. I have argued that ritual routinization helped the first farmers to settle down and create larger and larger cultural groups. But this was only the beginning of a much more complex process. As we close the current chapter on the Neolithic transition, we open a new one with an even more surprising storyline. This is a story in which routinization not only helped larger groups to form and stabilize but also to become dominated by leaders wielding previously unimaginable powers. Some of these new leaders were treated as gods; many established highly repressive regimes founded on slavery and human sacrifice. This is the point at which we leave the Neolithic behind and venture, with trepidation, into the Bronze Age.

5.

Religiosity and the Rise of Supernatural Authority

The biggest storage jars I have ever seen can be found at Knossos, Crete. Lying at the heart of what was once a grand palatial complex built around 4,000 years ago, these giant receptacles were designed to hold vast amounts of grain, wine, or oil. It has been estimated that something approaching 20,000 gallons of the stuff was poured into jars in the storage rooms of the palace.[1] The creators of these enormous containers, along with the surrounding buildings, public spaces, and artworks, belonged to a very sophisticated civilization. The Minoans had a highly specialized division of labour, and a very elaborate system of beliefs and rituals, magnificent architecture, works of art, and systems of trade and tribute linking them to other complex societies as far away as ancient Egypt.

This was a remarkable period in world history. Around the time that these Minoan palaces were being built, several other major centres of civilization around the world were also flowering – not only in the Mediterranean, Middle East, and North Africa but also in parts of India, China, and Peru. They began as small city states evolving into larger regional systems of governance. This same basic pattern unfolded in ancient Egypt and Mesopotamia, the Indus Valley in South Asia, the Yellow River Valley in the Far East, and the pre-Columbian civilizations of the Americas. In each of these regions, remarkably advanced societies evolved out of earlier Neolithic farming communities of the kind described in the last chapter.

Farming and the production of surpluses played a central role in this story. Starting more than 10,000 years ago, agriculture first appeared and spread from the regions in which the settlements at Göbekli and Çatalhöyük were established to engulf much of the Levant, Mesopotamia, and Egypt. Farming also spread east, as testified by important archaeological sites such as Mehrgarh in what is now Balochistan Province in Pakistan, which dates from around 8,000 years ago – bringing with it much the same package of practices associated with crop cultivation and animal domestication. Just like the peoples of Çatalhöyük, those at Mehrgarh built their houses out of mud bricks, growing wheat and barley in their fields and raising sheep, goats, and cattle. They also developed remarkable skills in the production of obsidian blades, arrowheads, mirrors, and jewellery. Such creativity became the hallmark of civilizations adopting agriculture. As populations grew larger, rates of technological innovation increased. Among these new inventions was the capacity to yoke oxen, horses, mules, donkeys, and water buffalo. The harnessing of traction animals in this way contributed to ever-increasing agricultural surpluses. And as societies grew wealthier, their trading networks expanded, and elites and hierarchies began to form.

In South Asia, farming societies evolved into the great urban centres of the Indus Valley from around 5,000 years ago. Parallel processes unfolded in China, eventually spreading major innovations in crop cultivation and animal domestication to Southeast Asia and from there across the Indian Ocean to Africa and likewise in the opposite direction, across the Pacific. And as farming methods travelled, they helped to fuel the rise of many of the world's most remarkable civilizations, from the kingdoms of Madagascar to the paramount chiefdoms of Polynesia. Moreover, independently of these events in Asia, Africa, and Oceania, a comparable evolutionary process unfolded, mainly starting around 2,000 years ago, in various parts of what is now the eastern USA, Mexico, and the northwestern coast of South America. In all those regions, advanced societies emerged – from the Mississippian cultures of Cahokia to the Mesoamerican cities of central America and the great Inca empire spanning most of the western coast of South America, including parts of the countries we now know as Peru, Ecuador, Bolivia, Argentina, Colombia, and Chile.

But for societies to grow in this way they needed a new form of social control: organized religion. While ritual routinization, discussed in the last chapter, may have been a necessary condition for the rise of more complex societies, it was not sufficient to support very large ones. From the dynasties of ancient Egypt to the paramount chiefdoms of Polynesia, larger societies depended on top-down systems of wealth extraction and political domination. And this took some doing. On the one hand, people had to be persuaded that unequal social systems were necessary and morally right – via the widespread adoption of shared norms, beliefs, and practices. On the other hand, mechanisms of enforcement were required to deter anyone who might be tempted to rebel. The interplay between persuasion and coercion took place largely in the domain of religious beliefs and rituals.

This was necessary because the effects of the conformism that we explored in the last chapter become weaker as populations become larger and denser. Part of the reason for this is that as we find ourselves spending more and more of our time interacting with relative strangers, the reputational costs of deviance are downgraded. People we barely know are in a weak position to tell us what to do when we step out of line. Hence the need for ways of actively deterring people from knowingly deviating from the script. Large societies developed a variety of new ways of policing conformism, all with one thing in common: they were closely associated with the exercise of supernatural power, harnessing humanity's natural religious predispositions and susceptibilities.

Exactly how power was exercised in ancient Crete remains something of a mystery. Some say that the enormous stockpiles of grain and other foodstuffs in Minoan palaces were part of a system of redistribution. Wandering around the ruins at Knossos, however, I found myself doubting this interpretation. As I stood gazing at the towering jars in the palace storage rooms, I began to suspect that something altogether more sinister was going on. Clearly this was a society with powerful royal lineages and wealthy elites. Their artworks suggested an admiration for highly controlled violence – from their exquisitely decorated daggers to sculptures and frescoes depicting male aggression, from boxing matches to images of men stabbing each other in the neck.[2] The pots seemed to me less like a product of redistribution than of expropriation.

The more I learned about the Minoan palaces, the more I began to suspect that behind the stylish tastes and lifestyles – seen everywhere from elaborate boardgames to massive serving dishes suitable only for the most lavish banquets – there must surely have been a darker side. In the ancient world, this kind of wealth typically came at a cost. Nobody really knows just how unequal or oppressive the society was but when one considers the extraordinary wealth, art, and food stores at Knossos, it is easy to imagine that this was a highly stratified society, in which the labour of many was exploited by a relatively small elite. I found myself wondering if the grand palaces of ancient Crete could have been built without slavery or at least some kind of large-scale exploitation of manual labour. Perhaps when the ancient script of the Minoan civilization is one day deciphered, we will have a clearer picture. However, there are other more direct signs of extreme inequality to consider. For example, there is evidence at a site close to Knossos of violent death consistent with human sacrifice.[3] The ritual killing of humans to please or placate the gods is one of the surest signs of top-down oppression that has ever been produced. Could it be that I was looking at a highly unequal social order, legitimated by appeal to the gods?

The main focus of this chapter is the part played by organized religion in the rise and spread of larger-scale, hierarchical, and centralized societies. As we saw in Chapter 2, our relationship with the supernatural typically takes a handful of extremely widespread forms – for example, to imagine that supernatural agents and forces can help or harm us, to assume that we live on after we die, to attribute intelligent design to the natural world, or to remember and pass on concepts that contradict our intuitions in ways that captivate and enthral. As societies grew larger and more complex, however, these features of wild religion were tamed and exploited in much the same way as traction animals were yoked together to pull a plough. Our religious biases were harnessed to control our behaviour. Eventually, our beliefs in the supernatural would compel us to submit to rulers, to pay them their dues, and fight their wars.

Three broad clusters of religious characteristics emerged over time, each exploiting our religious instincts in distinctive ways. First, our natural tendency to believe in an afterlife became a reason for expressions of obedience to capricious and often curmudgeonly ancestors. Next, our expectation that supernatural beings should be socially dominant became a way of legitimating novel forms of political domination. And then, finally, our moral instincts were harnessed in ways that enabled internally diverse empires, crosscut by differences of class and ethnicity, to cooperate and compete in an increasingly globalized world. This is not a simple story to tell, however: although these three dimensions of religion evolved in that order, they did not displace each other but often became overlain and interwoven. Between them, though, these three stages contributed immeasurably to religion as we now know it, forming rich seams in the faith traditions distributed across the world today.

Why worship the ancestors?

When followers of the Kivung in Papua New Guinea laid out food offerings in specially constructed temples, they hoped their ancestors would be pleased by the hard work and good intentions that went into the creation and presentation of their gifts. The goal was to create a feeling of obligation among the ancestors and a longing to be reunited with their descendants, thereby expediting the miraculous return of the dead that all Kivung followers earnestly anticipated. As we have seen, however, preparing and leaving offerings to please and placate the ancestors was an arduous business. Every day of the week for month after month and year after year

Kivung communities laid out offerings to the ancestors in the village cemetery temple. But they also laid out offerings in other kinds of temples. One of these was named after a prominent founder of the movement and needed to be tended twice a week. Another kind of temple, associated with individual families and their ancestors, also needed to be regularly tended. And in addition, all Kivung members needed to turn up at lengthy meetings resembling church services in a specially constructed village hall twice a week. Not to mention the many other rituals that needed to be conducted individually on a day-to-day basis as well as numerous abstentions associated with special holy days (two per week) and states of menstrual pollution (which typically affected at least some of the women in the village at any given time). In short, the need to satisfy the unremitting demands of the ancestors for constant displays of attention, devotion, and concern led to highly routinized forms of ritual in the Kivung.

All these beliefs had their foundations in panhuman religious intuitions concerning life after death. The Kivung concept of 'ancestor' was grounded in the idea that minds can be separated from bodies and can continue to live on even after the physical 'shell' has been buried in the ground and is rotting away – a belief that, as I explained earlier, is likely an expression of the way we naturally reason about mental states, and particularly higher-order cognitive functions such as memories, beliefs, and desires. But the Kivung harnessed these intuitions in new ways to motivate feelings of obligation to the ancestors that in turn motivated high levels of ritual conformism. What was salient about ancestor psychology was that it demanded fastidious respect, care, and attention from the living, such that adherents were compelled to carry out their duties with the utmost attention to detail, and that the ancestors would know and remember whether they had done so or had fallen short in some respect.

In addition, ancestors were attributed some mildly counter-intuitive properties, such as the ability to pass through walls (a breach of intuitive physics) and to see directly into people's hearts and minds (a breach of intuitive psychology). These were not just any old counter-intuitive properties, however; each had powerful social effects. They indicated that the ancestors not only expected their descendants to carry out their ritual duties conscientiously, but they could also check up on them, keeping the entire community under continuous surveillance by being able to move around at will and to see who was planning to cut corners or shirk their obligations. As such, Kivung beliefs in ancestors provided a formidable method of enforcing ritual routinization across the movement as a whole.

The desire to please and placate ancestors helped to transform followers of the Kivung into mass conformists, ensuring that the movement's beliefs and practices were maintained intact. But Kivung followers were not the first to stumble upon this method of enforcing conformity of belief and practice. Ancestors are the guardians of tradition in many regions of the world. Societies that venerate ancestors also respect their elders, who know the most about how things were done in the past. The elders are pleased when we follow their cherished traditions and personally offended and punitive when we don't. You could say that ancestors are elders writ large. They expect us to carry out all our ritual duties and observe the group's traditions to the letter – just like the elders – but they are more acutely offended than even the elders would be when we don't discharge our obligations. Moreover, they have greater powers to punish the lazy and disrespectful among us.

Ancestor worship, however, is not necessarily a moralizing form of religion because the dead often don't really care how we treat the living. We can be as deceitful, conniving, selfish, and aggressive as we like towards each other so long as we are respectful and dutiful towards the dear departed. And unlike moralizing gods imagined to be all-knowing, all-powerful, and always right, ancestors are commonly seen as fallible, capricious, and inconsistent. They might benignly overlook some ritual *faux pas* today and yet shower sickness and pestilence on you tomorrow for even the most minor of oversights.

The powerfully intuitive and socially useful nature of these ideas helps to explain their early emergence in world history. Indeed, they seem to have played an increasingly important role in the way group identities became established in the Neolithic. This is to be seen in the archaeological evidence for a growing preoccupation with secondary burial practices as farming societies grew in size, suggesting an increasing preoccupation with the propitiation of ancestors. Whereas primary burials are occasions for the interment of corpses, usually soon after death, secondary burials involve the removal of remains much later – sometimes years later – for the purposes of further ritual treatment. The way this was carried out in ancient prehistory varied enormously across different cultural groups – ranging from the stripping away of soft tissue from the human skeleton using blades through to the putrefaction of the corpse to 'de-flesh' the bones more slowly. In some cases, it was not the skeletal remains that were venerated but the ashes of incinerated corpses or other durable objects associated with the dead. Whatever form such actions took, however, they always seem to have been about creating an enduring social relationship with the dead, via some part of their remains that could be preserved, providing a point of connection with the everlasting spirit.

Secondary burial practices are remarkably widespread in the human past. From skull cults at some of the very earliest Neolithic sites in Palestine to the ancient tombs of Ireland, and from the funerary enclosures of northern Israel to the cemeteries of ancient China, the world's earliest farmers were clearly very concerned about maintaining relationships with their dead.[4] Bones and stones are often all that remain of these ancient societies, but the bones in particular reveal a seemingly endless variety of ways of disposing of corpses, such as de-fleshing and re-fleshing them, removing heads, jaw bones or other body parts and carefully placing them in significant locations, dressing and undressing the skeleton, presenting them with offerings, painting them in ochre, or placing shells in their eye sockets. Consider the houses – especially the history houses – of Çatalhöyük that we encountered in the last chapter, in which the dead were buried and reburied repeatedly over the course of multiple generations.

This desire to connect with the ancestors, even those who died before anyone handling their bones had been born, suggests a growing concern with the group's origins: it seems clear that early farmers felt strong obligations to their ancestors. It therefore seems highly plausible that this would have acted as an enforcing mechanism when it came to the faithful performance of rituals and the maintenance of revered traditions.[5] But what is particularly striking is the extent to which this shift is associated with the transition to agriculture. Although the details of how ancestor worship was conducted and conceptualized remain obscure, analysis of the behaviour of traditional societies surviving today suggests that the increasing role of the dead in ancient societies was closely related to the rise of farming. Using large samples of ethnographic data on modern societies to infer the relationship between agriculture and ancestor worship during the Neolithic, and deploying sophisticated statistical techniques to reconstruct the evolutionary trajectories of ancient societies, researchers have argued that the veneration of ancestors really took off with the intensification of farming practices.[6] The link between ancestor worship and agriculture may be observed also in the ethnographic record. For example, one study comparing 114 societies has shown that hunter-gatherers are the least likely to worship their ancestors, subsistence farmers more so, and advanced agrarians most of all.[7]

This may partly be explained by the usefulness of ancestors in sorting out legal matters. After all, claims to ownership of agricultural land and other resources are commonly based on prior possession or use, demonstrated with reference to traceable ancestry and its physical proofs. In traditional land disputes the world over, opposing sides appeal to principles of descent and inheritance. Maintaining close relationships with the dead is therefore important and the best way to show it is through the observance of ritual obligations. But in addition to this, beliefs in demanding ancestors would likely have motivated people to fulfil their obligations to living kinsmen and in-laws. This would have been particularly crucial in the wake of the transition to agriculture, because such duties were necessary for the production of surpluses – which, as we saw in the last chapter, were vital to survival in a household-based economy.

The veneration of ancestors had ramifications well beyond the realms of economics and law, however. In offering a supernaturally sanctioned method of policing adherence to the group's rituals and norms, ancestor worship helped standardize and preserve a society's cultural traditions, in turn allowing it to stabilize and grow. But in time, that would bring unexpected consequences. At first, ancestor worship would have allowed relatively egalitarian social institutions to flourish. For several thousand years after the invention of farming, communities grew larger and denser but not much more hierarchical. Yet as farming intensified and surpluses became more abundant, ancestor worship turned into something much darker. Instead of simply obeying rules that were accepted by all and enforced via peer pressure, ideas about the appeasement of ancestors came to be associated with increasingly oppressive forms of top-down enforcement.

Before that could happen, though, a fundamental change needed to occur in the way people thought about leadership. We needed to make the leap from simply respecting people because of their achievements to bowing down to them because of supernatural qualities and powers thought to be inherited by birth. This way of thinking eventually led to the sanctification of elites and the deification of rulers.

Divine sanction and worldly authority

Throughout world history, leaders have with astonishing frequency claimed to speak on behalf of the ancestors, gods, or any number of other supernatural beings. On the face of it, this is a peculiar phenomenon; it's difficult to imagine how any individual could stand up and claim the patronage of the supernatural without being very rapidly ridiculed or cut down to size. So how did chiefs, kings, and emperors establish themselves as the instruments for higher powers?

Anthropologists have described this transition in Pacific Island societies using the distinction between 'big men' systems and chiefdoms.[8] Big men achieve their positions of power and influence through personal skill and effort, winning over a coterie of followers and offering them protection in return for support and labour. For example, the term 'big man' is often used to describe a common form of leadership in Melanesian societies, whereby individuals gain influence and respect by demonstrating qualities that are much admired: bravery, industriousness, generosity, oratorical skills, magical powers, and the ability to amass wealth.[9] But crucially, when the big man dies or loses his grip, his authority dies with him. By contrast the status, power, and wealth of the chief is largely ascribed, typically as a birth right. Often it takes the form of some mystical essence, passed down through the chiefly lineage.

Inherited chiefly rank provided a foundation for something even more revolutionary: the emergence of divinely sanctioned kingship in much larger states and empires. In many of these social systems, the dominion of chiefs gradually expanded, with rulers taking on ever more formidable powers. Kings and emperors in regions as far apart as South America, Africa, and Asia were eventually divinized or at least seen as wielding supernatural powers. In his grand comparative study of early civilizations, archaeologist Bruce Trigger traced the rise of divine kingship in regions as far flung as the Americas (Aztec, Inca, and Maya cases), Africa (Egypt and Yoruba), Mesopotamia (Akkadian empire), and northern China (Shang).[10] Kingship in such societies, as Trigger explained, was underwritten by supernatural power – wielded ostensibly for the protection of everyone from natural disasters but also used to justify the punishment of individuals foolish enough to display a lack of extreme deference, typically by putting offenders to death. In states where kings were treated as godlike, such as Hawaii, subjects had to prostrate themselves in the presence of a monarch, while in Benin merely suggesting that the king had human needs (e.g., to eat or sleep) carried the death penalty. But even where kings were not deified, they were thought to wield supernatural powers. For example, Aztec kings – though not gods – were dressed in the regalia of deities during inaugurations. Likewise, early dynastic rulers in Mesopotamia bore regalia believed to be of divine origin and the institution of the kingship was considered to be god-given.[11]

Once established, systems of kingship have come to hold great psychological appeal – perhaps because they are supported by our most basic intuitions about leadership. We saw in Chapter 2 that even before they can talk, babies expect supernatural beings to be socially dominant. This suggests that humans everywhere naturally expect individuals who exhibit minimally counter-intuitive powers – such as the ability to perform magic, foresee the future, commune with the dead, or divine the causes of misfortune – to command the respect of lesser mortals. In ancient foraging societies, such individuals might have been shamans, witchdoctors, or masters of initiation. But with the transition from big-man societies to larger and more centralized institutional systems, leaders were increasingly imagined to wield much mightier forms of supernatural power.

Nevertheless, while it may be easy to understand how we end up accepting such leaders once they are well established and surrounded by powerful armies and imposing fortifications, that doesn't tell us much about how such systems got off the ground in the first place. While it is hard to reconstruct in detail how self-made leaders turned into systems of inherited rank for the first time, in various societies we can say with some confidence when it first happened. In ancient Egypt this leap occurred thousands of years ago whereas in sub-Saharan Africa it occurred much later. In some regions, we might venture to pinpoint when the threshold was passed. For example, Naram-Sin, who reigned over the Akkadian empire from 2254 to 2218 BCE, was the first ruler in Mesopotamia known to have been attributed divine status.[12] He not only persuaded the people of Akkad to accept that he was a living god but also that he could pass on his divine essence to his offspring – indeed his daughter became the only known example of a woman and non-king to be divinized in this ancient civilization.

The first societies to make this leap from 'big men' to hereditary chiefs probably did so in fits and starts. We have examples of societies that made it across the threshold, only to collapse again, their fate being recorded by explorers, missionaries, or invaders. In some places, we know this happened not once but multiple times. For example, among the hill tribes of highland Myanmar (formerly Burma), descent groups experiencing exceptionally good harvests were occasionally able to convince their neighbours that they were descended from powerful spirits, thereby legitimating claims to inherited rank. But their efforts to imitate the hierarchical Shan state system invariably failed and claims to chiefly rank collapsed. It has been argued that this pattern repeated itself many times among groups such as the Kachin, Naga, Wa, and Chin.[13]

So, making the transition is not easy. But we can gain a glimpse of how leaders try – and sometimes fail – to achieve it by looking to the ethnographic record. First, that involves recognizing that not all types of supernatural leadership are the same. Many types of leadership involving the exercise of supernatural powers can be placed on a continuum, from the lowliest to the most exalted.[14] At the lowliest end lies the victim of spirit possession, whose body is invaded by the unbidden presence of an alien being. The person is helplessly taken over and unable to resist. Such individuals may, in a state of possession, have influence over their community but since they are thought to exercise little control over the experience themselves, they are not accorded much respect or authority as leaders. By contrast, a spirit medium is thought to have more power over the situation. Mediums are typically seen as regulating the entry and exit of spirits and, as such, accorded some authority over them and over their fellow mortals. Further still along this continuum is the prophet, who not only exercises control but acts as an interpreter of the forces from the 'other side'. Such individuals wield higher authority in themselves, by not only channelling the spiritual world but shaping its impact on this world. And finally, at the most exalted end of the spectrum lies divine kingship and apotheosis. Absolute power on this earth lies in the absorption of the spirit world, in which the leader becomes a pure embodiment of the supernatural – not a fleeting vessel but a permanent envelope for the divine. Such figures of authority – whether in the form of emperors or messiahs – are seen as infallible. Their word can easily become law.

This scheme captures some key features of the way leadership resting on claims to supernatural power has evolved over thousands of years. But there are also rare moments when anthropologists have been able to observe the dynamics of the transition in real time. I was fortunate enough to witness one of these extraordinary transitional processes unfold before me, during my fieldwork among the Baining in Papua New Guinea. Traditionally, the Baining did not have hereditary chiefs. Positions of leadership in the community had to be achieved by gaining a reputation as a pillar of the community. For a man, this meant accomplishing great feats as a warrior, orator, magician, or ritual expert. For a woman, it typically meant bearing and raising many children or taking a leading role in the secret initiations of girls or the organization of communal feasts. But nobody claimed the right to boss other people around based on claims to sacred authority. Leaders lacked the spiritual superiority of inherited rank and there were no chiefs or kings who claimed to be gods or even to be acting on their behalf. But during my fieldwork, I had an exceptionally rare opportunity to witness efforts to change all that.

In my village, a young man in his early twenties – Tanotka – suffered a delirium. This was probably brought on by cerebral malaria, which was a common and often deadly affliction in the village.[15] However, the strange utterances he produced during his illness prompted many to conclude that he was undergoing possession by an ancestor. One of the most talked-about claims made by Tanotka during his feverish state (or rather by the ancestor speaking through him) was the statement: 'I am a post.' According to his elder brother – a man known as Baninge – this was a reference to the central post used in traditional Baining houses to support the roof. These roundhouses had conical roofs constructed around a ring of posts, the rafters converging on a central post in the middle. According to Baninge, the ancestor speaking was saying that he would support the community (symbolized by the ring of posts) so that they could lean on him (like the rafters leaning on the central post) in their efforts to be reunited with all their dead relatives. This was taken to be a sign that the miracle of returning ancestors was now imminent.

To translate these elements of leadership into the scheme described above, we might say that Tanotka was merely a helpless victim of spirit possession, unable to control his predicament. But his brother, Baninge, assumed a leadership role more like that of a spirit medium, channelling Tanotka's message. More than that, however, Baninge wanted to serve as a guide to the will of the ancestors. In fact, after Tanotka recovered from his illness, he started to have further visions and to make additional cryptic statements, working closely with Baninge to make sense of it all. Together, they assumed the role of prophets – interpreting messages from the other side. Tanotka gradually began to withdraw from public life, becoming a recluse, and those around him argued that he was turning into an ancestor – a kind of god incarnate. Little by little, Tanotka was undergoing an apotheosis, soon to emerge at the far end of our continuum as an envelope for the divine – a Baining messiah. Unlike a big man, whose authority is expunged upon his death, Tanotka was turning into a living divinity. As he came to be seen increasingly as a living god, he no longer needed to appear in public or to do anything of social significance. If he had been able to consolidate this position and sire a son, there would surely have been plausible grounds to claim that Tanotka's offspring would inherit his divine status.

Tanotka's career exhibited all the hallmarks of the transition from the individual 'big man' to the hereditary 'chief'. But it was not to be. The transition from mediumship to divinity is one that has been attempted innumerable times in human societies on the cusp of hierarchy, but it is not easy to pull off. The main reason, as Tanotka was to discover, is that claims to divine status attract scepticism and critique. Humans are sensitive to cues that claims to supernatural status are self-serving, legitimating claims to social dominance. For this reason, Tanotka could not declare himself to be a living ancestor but could only aspire to that position via the efforts of another. If Baninge pushed him forward, then Tanotka could feign reluctance to assume the mantle of authority before eventually acceding to his exalted destiny. Despite this Machiavellian ingenuity, however, Tanotka's claims to leadership eventually collapsed. In part, this was because the splinter group with which Tanotka was associated was forced to disband following the failure of his prophecies. His followers destroyed all their livestock and consumed the meat in lavish feasts to celebrate the return of their ancestors. But when the ancestors never arrived, they were faced with the threat of starvation and had to give up on the splinter group and its leaders, returning to the drudgery of everyday life.

It is clear that efforts to achieve chiefly or aristocratic status based on claims to supernatural power did not meet with widespread or lasting success in the traditional cultures of mainland Papua New Guinea. But in the example of Tanotka, we get a glimpse of how one individual might – in the right circumstances, if making the right claims, and being supported by the right backers – lay claim to supernatural authority. This must have happened also in the prehistory of civilizations ranging from ancient Mesopotamia to the Inca empire. But one need not travel too far from mainland Papua New Guinea to find examples of leaders with better luck than Tanotka laying claim to this kind of power. Analysis of ancient DNA has shown that the ancestors of Tanotka migrated to regions of the Pacific where chiefdoms did indeed evolve, whether independently or through the spread of customs from other islands, such as those in the Polynesian Triangle between New Zealand, Hawaii, and Easter Island.[16] Such societies developed well-established hierarchies and leaders who derived their authority from a heritable form of supernatural power known as *mana*. Some of these chiefs ruled over entire archipelagos, passing on their dominions to their sons and heirs. Those with *mana* did not simply grab power, they came to embody it.

Once this transition has been accomplished, its effects are transformative. Within a few short generations, heritable leadership can become firmly rooted in tradition. And that leads to expansion: if leaders can pass on worldly authority to their children, then their factions are no longer destined to implode after they die – instead, they can grow over successive lifespans through systems of inheritance. This invention therefore marks the shift from a relatively small-scale social system, based on a balance of power among groups formed through bonds of kinship and descent, to the first much larger-scale, hierarchical societies with ruling elites.

For the early farmers of prehistory, this would prove to be a mixed blessing. Once discovered, heritable forms of supernaturally sanctioned power revolutionized the political landscape, creating wealth and power – but not for everybody. This turn had a dark side. The earliest rulers wielding supernatural power tended to take the idea of political domination to extremes. Nowhere is this more evident than in the widespread practice of human sacrifice.

The rise of human sacrifice

The Aztecs of central Mexico ripped out the hearts of countless sacrificial victims to feed the rapacious appetite of the sun god, lest he should weaken and be unable to make the daily journey across the sky. But they also sacrificed humans to celebrate communal achievements, such as the completion of temple-building projects. Though unimaginably painful, this was not the worst possible way to die.

Some Mesoamerican sacrifices involved protracted torture prior to the extraction of the heart, such as the flaying of victims and various forms of bloodletting.[17] For the Inca, a common method of sacrificial killing was burial alive.

Some of what is known about these practices has been gleaned from the writings of Spanish chroniclers, which require considerable expertise to interpret correctly with minimal risk of undetected bias or exaggeration. But some of the evidence survived into modern times because the bodies of sacrificial victims were thoroughly preserved. For example, in the century prior to Spanish conquest in 1532 CE, the Inca built shrines on top of mountains and left the bodies of many sacrificial victims to freeze, only to be recovered in the twentieth century, when their mummified corpses were still in remarkably good condition.[18]

Based on a combination of written records and autopsies, researchers can be confident that large numbers of humans were sacrificially killed by the Inca and for a great variety of reasons, such as to celebrate the lives of emperors, to ensure fertility and health, or to seek atonement in response to natural disasters. Many of those put to death in this way were children – aged between four and ten years in the case of boys, while girls were usually kept alive into puberty because female virginity was much prized in sacrificial victims. The scale on which such killings were carried out is staggering to contemplate. In one case, around 200 children are believed to have been sacrificed to celebrate the coronation of a single emperor.

The terror and agony of sacrificial victims in such societies is hard to get our heads around today. But it is also very troubling to contemplate the suffering of their loved ones. What did it feel like as a parent to have to give up your beloved offspring to be tortured and killed? It is hard to imagine a more extreme expression of the power of rulers. Among even the most repressive of political regimes present in the world today – including those that carry out public executions or engage in the torture of dissenters – human rights organizations would be hard-pressed to identify forms of state violence that are ghastlier than human sacrifice as practised in past ages. And yet, although human sacrifice is now extremely rare, it has not always been that way. Indeed, throughout the earliest chiefdoms and states, from western Eurasia to sub-Saharan Africa, from Polynesia to the Americas, and from northern Europe to East Asia, human sacrifice was remarkably widespread. Why?

One explanation for the rise and spread of human sacrifice is that it was a successful cultural adaptation, allowing the groups adopting it to become larger, wealthier, and better organized, enabling them to spread and to conquer their rivals. If you had been born into such a society, you might consider yourself fortunate in many ways. But, on the other hand, unless you were among the ruling elite, you might also be inclined to think that the system was rather unfair. Power holders were capable of surrounding themselves with levels of wealth and luxury denied to the common people. Not only were power, wealth, and status concentrated in elite groups but members of such groups typically lived a life of relative leisure – benefiting from the fruits of other people's labour. And they did so by claiming that it was all part of a divine order. Religion became a basis for legitimation of inequality. More than 5,000 years ago, Sumerian city states were already being ruled by priestly governors or by kings associated with the city's religious rites and patron deities. Not much later, Egypt was unified under powerful kingdoms by pharaohs legitimated by supernatural powers whose rituals demanded the taking of human lives. Similar patterns unfolded in South Asia, China, Mesoamerica, Polynesia, and elsewhere.

In many ways such unequal political systems go against our egalitarian impulses. How did people come to accept that elites contributing so little to the collective good should be allowed to enjoy so many privileges? The answer lay in religion. The gods demanded that we accept extreme forms of inequality. And nowhere was this dogma more strikingly expressed than in the institution of human sacrifice. Elites that were able to use institutionalized violence to strike terror into the general populace were better able to sustain extreme inequality and therefore to expand their dominions through the exercise of brute force. Human sacrifice is just one instance of this general approach to empire-building and went hand in hand with other extreme forms of domination, from enslavement to brutal forms of colonization, torture, and public executions. But while all these forms of repression appealed to cosmic forces, human sacrifice did so in the most striking way possible. It was not merely endorsed by the gods – it was demanded by them.

We understand this phenomenon better than ever thanks to the wondrous diversity of Pacific Island societies, which provide a suitable natural laboratory and testing ground for our theories.[19] Austronesian-speaking peoples originated in what is now Taiwan but their seafaring descendants spread across vast areas of the Pacific and Indian oceans, colonizing islands as far apart as Polynesia and Madagascar. Many of the societies they spawned crossed the threshold of heritable authority – establishing a great diversity of aristocratic and royal lineages, basing their right to govern on claims to supernatural power. Human sacrifice was widely practised in these societies, despite the fact that their cultural belief systems varied enormously and the methods of killing sacrificial victims were similarly diverse – from bludgeoning, crushing, decapitation, and cutting into pieces through to strangulation and drowning. These varied forms of ritualized killing were typically carried out on the command of elites – priesthoods, chiefs, and kings – whereas the victims were often of lowly status (e.g., slaves, children, or captured enemies).

To investigate the relationship between social inequality and human sacrifice, researchers began by assembling a sample of ninety-three Austronesian language groups. Those lacking mechanisms for the inheritance of wealth and status – based on the big-man systems I described earlier – were classed as 'egalitarian'. Those that had made it over the threshold of establishing systems of inheritance, but which still allowed upward mobility within a single generation, were classed as 'moderately stratified'. And those that had heritable differences of wealth and status but little opportunity for upward mobility within a generation, were classed as 'highly stratified' – meaning, basically, that positions in the social order were pretty much ascribed by birth. The study found that human sacrifice was most widespread in the highly stratified societies but rare in the egalitarian ones. Using language trees to plot how the various Austronesian-speaking groups had evolved over time, the researchers demonstrated that the adoption of human sacrifice by any given group in the past increased the chances that it would become highly stratified in the future and reduced the chances of stratification disappearing once it had become established. In other words, there was a strong relationship between human sacrifice and social inequality – consistent with the idea that human sacrifice came about as a way of enforcing the obedience of populations by instilling fear.

The institution of human sacrifice wasn't just a demonstration of brute force, however. It was also a sacred duty and a holy contract between humans and the gods. The Inca empire spanned a region of South America where tectonic plates collided, causing frequent and catastrophic earthquakes and volcanic eruptions. Efforts to please and placate the deities thought to be responsible for these natural disasters focused on costly gifts. And what could be more precious than one's own beloved children? To the extent that Inca subjects bought into this system, they probably didn't live in a state of resentment at the injustice of it all but rather accepted it as part of the given order of things. Although some may have been tempted to rebel, most people were likely to have been quiescent. The same was no doubt true in Polynesia and in many other political systems that practised human sacrifice. For example, cultural groups as diverse as the Mongols, Celts, Scythians, and early Egyptians all sacrificed humans at one time but it is unlikely that people in these societies generally regarded it as a form of cynical oppression. In all the early civilizations where humans were killed to please or placate the gods, most people likely regarded such practices as a necessary evil.

This all raises a question: even if people did not actively resent the acts of extreme violence going on all around them, how could they work as a way of instilling social control? The answer can perhaps be found in what psychologists call terror management theory, or, more usually, 'TMT'. TMT suggests that human sacrifice and other grisly religious practices that make death and dying more salient, would have instilled fears of mortality and feelings of existential anxiety that in turn contributed to conformism – and so benefited ruling elites, who wanted people to go along with the existing order of things. The basic idea behind TMT is that the ability of humans to recognize their own inevitable demise – the certain knowledge that we will all die one day – directly conflicts with our evolved urges to survive and reproduce. Thanks to this clash between the drive to cling onto life and the awareness of our own mortality, we are all condemned to live in a state of anxiety. Advocates of TMT argue that when one's own mortality is made salient, whether consciously or unconsciously, people adopt a variety of defensive strategies designed to reduce negative stress.[20] One such way is to create the fantasy of eternal life, and much research inspired by TMT suggests that afterlife beliefs, as elaborated by religious traditions, function to relieve anxieties about death. The idea is that people become more religiously committed whenever the prospect of death is made more real or immediate. This is reminiscent of the old adage that there are 'no atheists in foxholes' – meaning that soldiers risking death on the frontlines are more likely to believe in a deity and an afterlife than people in more secure and unthreatening environments.

The very existence of religious systems focusing on death and the afterlife is itself a way to get people to think about their mortality (and so defer to norms and authority figures). Anthropologists have often made the point that ancestor worship makes the theme of death far more salient than it would otherwise be – in particular, by focusing attention on the treatment of corpses in various kinds of rituals but also on cemeteries, tombs, and other sites of interment.[21] These morbid preoccupations were arguably taken to a new level by the practice of human sacrifice. Indeed, it is hard to imagine a more extreme example of how religion might evoke a sense of terror at the prospect of death than one in which human beings are gruesomely killed in public to please or placate a deity. In fact, concerns with the dead and human sacrifice are often combined. Although the Inca are most famous for offering up human lives to the gods, they were just as remarkable for the depth of their genealogies, tracing back relations with the dead up to twelve generations.[22] If the TMT evidence is to be believed, this pervasive emphasis on dying and the dead arguably increased levels of deference to authority and obedient adherence to norms.

All this seems to imply that human prehistory is marked by a long, gradual progression through increasingly coercive mechanisms of religious norm enforcement. It all started innocuously enough. The first farming societies managed to grow larger by generating surpluses that were sufficient to see them through hard times, and in particular to subsidize households going through particularly difficult phases, for example when there were many mouths to feed relative to active pairs of hands. Ancestor worship helped to push people to meet their obligations to needy relatives and not simply to put their feet up when their own subsistence needs were satisfied – by instilling a profound and abiding respect for one's elders, whose incessant demands must be met even from beyond the grave. But as societies grew larger still, this mechanism alone was not enough to motivate hard work and devotion. The rise of hierarchies and elites meant yet more relatively unproductive consumers to support – not just children too young to shoulder adult burdens or elders too frail to labour but professionalized priesthoods, aristocrats, and rulers who expected to be waited on hand and foot. To motivate that kind of obedience required more than merely devotion to beloved family members. It required fear. This is what many of the religions of early states offered in spades: from a focus on bloody warfare and the taking of slaves to the practice of human sacrifice.

The demise of human sacrifice

Nevertheless, if human sacrifice was a mechanism of social control, becoming more prevalent as ancient societies became increasingly hierarchical and economically unequal, then why isn't it still widespread in the world today? To answer that question, we needed a way to rigorously compare human societies around the world over multiple millennia – not an easy feat. Answering questions about general trends and causal arrows in world history is notoriously difficult. Perhaps the biggest problem of all is selection bias – the risk of picking examples from past societies that fit our theories and overlooking evidence that doesn't. What we require is a more objective way of establishing and explaining patterns in the evolution of human societies and cultures – a method that will allow us to quantify those patterns and test theories of causality without knowing in advance what the results will look like and without being able to influence the outcomes in ways that could favour our preferred accounts.

This idea led me to Peter Turchin – one of the world's most productive polymaths who shared my fascination with explaining how we got here. Together we developed a vision for the largest databank of world history ever constructed. Using one of my research grants, we hired a postdoc – Pieter François – to help us get things off the ground and together we named our project after the ancient Egyptian goddess of writing and record-keeping: *Seshat*.[23] The goal of Seshat: Global History Databank is to test theories about the way human societies have evolved by assembling a vast amount of information on world history and analysing it statistically. One of our topmost priorities was to establish the role of religion in the evolution of social complexity.

Comparing information about the history of civilizations as far apart as Rome and Hawaii, however, requires a common set of variables. Unfortunately, the writings of historians, anthropologists, and archaeologists do not provide that kind of information in an easily accessible and analysable form. To facilitate comparison, we needed to create a code book, listing all the variables of interest, so that we could then investigate with the help of experts whether or not those specific variables were present or absent in each past society. Our goal was to capture as much diversity in global history as possible and to do so in a continuous, unbroken time series for a period of up to 10,000 years, based on scholarly sources.

To gather and organize published data in this way required an army of research assistants, methodically going through books and journal articles and extracting the information required, as well as countless hours of advice and input from scores of professional historians, archaeologists, and classicists to check the work of the research assistants and help to improve upon it. Our starting point was to decide which parts of the world to cover. Although the aim is for Seshat eventually to cover every square mile of the inhabited world, we realized from the outset that this would be impossible to achieve in the first ten years or so. In order to be able to get the project off the ground and show what it could do, we needed to begin with a reasonably representative slice of world history. Since our main interest was in the drivers of socio-political complexity, we needed a sample of the world's political systems that captured as much of the range of different types and stages of socio-political evolution as would be manageable given the resources at our disposal.

We started by dividing the world into ten regions, before looking for three 'natural geographic areas' (NGAs) of roughly 100 square kilometres (40 square miles) within each – typically a river valley or a plateau or some other ancient and naturally occurring feature that tended to retain its distinctiveness over the timescales of thousands of years. The aim was to maximize diversity in the sampling scheme by ensuring that for each world region, we could have at least one NGA in which social complexity emerged relatively early (shown on this map as a large circle), one relatively late (small circle), and one somewhere in the middle (medium circle).[24]

These dots constituted our sample of thirty NGAs from which polities were selected.

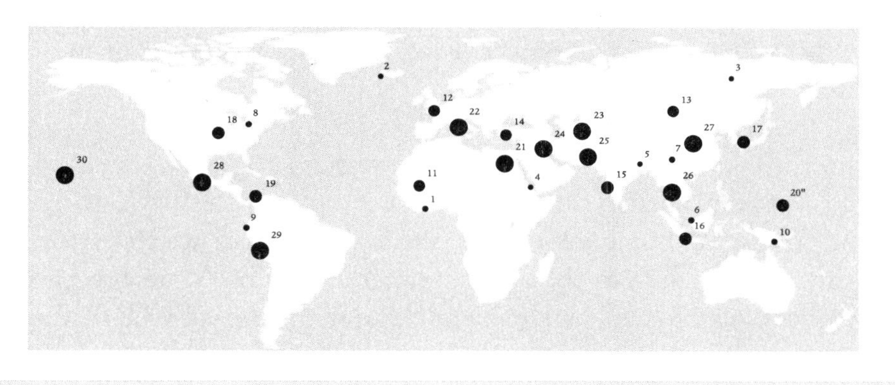

Vorld Region	Late appearance of complex polities	Intermediate appearance of complex polities	Early appearance of complex polities
.frica	1-Ghanaian Coast	11-Niger Inland Delta	21-Upper Egypt
urope	2-Iceland	12-Paris Basin	22-Latum
:entral Eurasia	3-Lena River Valley	12-Orkhon Valley	23-Sogdiana
outhwest Asia	4-Yemeni Coastal Plain	14-Konya Plain	24-Susiana
outh Asia	5-Garo Hills	15-Deccan	25-Kachi Plain
outheast Asia	6-Kapuasi Basin	16-Central Java	26-Cambodian Basin
ast Asia	7-Southern China Hills	17-Kansai	27-Middle Yellow River Valley
lorth America	8-Finger Lakes	18-Cahokia	28-Valley of Oaxaca
outh America	9-Lowland Andes	19-North Colombia	29-Cuzco
)ceania-.ustralia	10-Oro, PNG	20-Chuuk Islands	30-Big Island Hawaii

Source: The Seshat Team

Within each of our thirty NGAs, distributed around the world, our aim was to trace the evolution of polities from the earliest times on which data was available up to the arrival of industrialization.

This would allow us to track the evolution of social complexity over the grand sweep of world history, from the end of the Palaeolithic era when all humans lived in small foraging bands to the point at which most of humanity was living in centralized states of more than a million individuals.

To accomplish this, however, we needed to establish a way of measuring that elusive dimension we were calling 'social complexity'. As I explained in Chapter 4, social complexity is partly a matter of group size. But it is also about the scale on which we cooperate, and this is related to the structure of our groups. More complex societies tend to have a more hierarchical structure, with the institutions of government centralized in capital cities and overseen not only by powerful rulers but also administrators, priesthoods, military leaders, and other elite office holders. Scholars have long argued over which features of social complexity are most important in understanding social evolution. Rather than taking sides in these debates, the Seshat team took the most prominent features that scholars claimed to be important and coded for every single one of them. This produced a list of fifty-one main variables.[25] We then grouped our fifty-one main features of social complexity under nine main headings: population, hierarchy, government, territory, infrastructure, money system, literacy, information system, and size of the capital.[26]

Having coded for all these variables in hundreds of societies going back over thousands of years of world history we discovered something quite remarkable. Statistical analysis of the relationships between these variables showed that they were all much more closely associated than any of us imagined. For decades, scholars had been arguing over which different components – from territory size to number of levels in the hierarchy – should be the key metrics for measuring social complexity. Our method concluded the answer was: all of them. Where every scholar had made their mistake was in arguing that their rivals were wrong. All fifty-one features move together as social complexity increases.

With this insight in hand, we were finally getting somewhere on the question about the rise and fall of human sacrifice. We set out to use Seshat's database to establish how human sacrifice and extreme deference to authority related to the growth of social complexity. As we suspected, Seshat data confirmed that human sacrifice correlated with the initial rise of socio-political complexity. But eventually, and rather remarkably, human sacrifice disappeared in each of our regions. One hypothesis is that beyond a certain threshold in the evolution of social complexity, extreme forms of inequality and dictatorship cease to be viable.

Why? A clue would seem to lie in the way human sacrifice was woven into the cultural fabric of the states in which it was practised. States adopting this grisly practice typically had a shared cultural system in which the relationship between elites and commoners was characterized by reciprocity and not simply naked coercion. This can clearly be seen, for example, in the way Inca elites sponsored local festivals, receiving in return tribute, labour, and military support.[27] State provision in this way helped to ensure not only the perception of reciprocity between rich and poor but also – and most importantly – a set of shared beliefs and practices legitimating the cosmological system in which human sacrifice featured. As a result, human sacrifice and other strategies of rule by terror would seem to work best in highly homogeneous societies, where most of the population shares a common sense of loyalty to each other and to the social order. The more culturally homogenous the state, the easier it is to legitimate the exercise of extreme power and top-down coercion.

By contrast, extreme inequality was much harder to pull off in multi-ethnic empires composed of culturally diverse populations. When empires become more internally differentiated, incorporating distinct ethnicities, religious sects, professional guilds, classes, and other interest groups, extreme inequality in the form of human sacrifice and mass slavery seems to become increasingly difficult to maintain.[28] In other words, very unequal societies can be brittle, especially when their citizens become more diverse, making them vulnerable to internal dissent and defeat on the battlefield.

The chequered history of China provides many illustrations of this point. For example, when the Zhou dynasty overthrew the Shang around 3,000 years ago, the ranks of the former were greatly swelled by slaves from the latter, seeking liberation from their oppressors. Armies often recruit the oppressed and this is one of the prices that exploitative regimes have been forced to pay in the rise and fall of civilizations. Much later still, nearly a thousand years after the establishment of the Zhou, China was united on an even larger scale by the all-encompassing Qin Dynasty. However, this involved systematically undermining the basis for national identities among the many formerly warring states – including the burning of books and the execution of academics (a salutary lesson for people in my own profession). The use of such oppressive methods created only a tenuous and short-lived peace. Within just a few years of its establishment, the Qin ascendancy was replaced by the Han Dynasty.

Another classic illustration of this problem would be the repeated secessions of plebeian soldiers which began in the early Roman Republic, as it was repeatedly attacked by warlike Volsci, Sabines, and Aequi. Keeping the military on side proved possible only by making repeated concessions, eventually leading to major improvements in levels of social equality.[29]

The key point is that large, internally diverse empires cannot hope to persist if they rely too much on coercive tactics. It is difficult to specify precisely the moment at which societies reach the point when despotic control ceases to be enough to hold them together. However, our analyses of Seshat data suggest it occurs somewhere around the stage at which the overall population rises above a million individuals. Ancient societies crossing that transition point typically did not get that big simply by having high birth rates. They usually grew to such gargantuan proportions by absorbing other groups, often by means of conquest and invasion. However, large and unwieldy empires, composed of multiple ethnic groups, are vulnerable to revolts and revolutions in their hinterlands, fuelled by feelings of injustice and historic grievance among vanquished tribes. The examples of China and Rome above are not alone; this is a recurrent feature of the rise and fall of civilizations more generally. One obvious solution to internal dissent was to raise larger armies to put down any potential unrest – but that would require ways of inspiring loyalty among groups originating far from the centres of power. Ultimately, multi-ethnic empires needed a way of unifying their increasingly diverse subjects behind a set of common ideas. But how?

The answer to this question probably kept many ancient emperors awake at night. However, some of them eventually stumbled upon a solution – one that required the role of religion in political life to change again. For the first time, the gods began to develop a conscience.

Moralizing religions

The first large-scale moralizing religions – forerunners of the world religions we know today – flowed down two mighty rivers of religious revelation.[30] The source of one of these rivers might be traced to the Egyptian goddess Ma'at – arguably the earliest moralizing deity ever worshipped, dating back to the middle of the Old Kingdom of Egypt some 4,500 years ago. Ma'at personified various cardinal principles associated with justice, law, order, and balance. However, since Judaic prophetism was in part a reaction against the rulers of ancient Egypt, we might instead regard this tradition as the source of the western flow of more egalitarian philosophies or at least one of its major tributaries. Meanwhile, from the east came a set of revelations disseminated by Siddhartha Gautama during his wanderings through the lower Indo-Gangetic plain. Buddhism, as it became known, emphasized values of restraint and kindness, ideas patently uncongenial to the goals of despotic rulers and military leaders. In the confluence of these two mighty rivers of ideas, a host of new religions emerged during the first and second millennia BCE, among them Zoroastrianism, Judaism, and Jainism. These kinds of new religions then spread even further east – Buddhism reaching China during the Han period and later Manichaeism. This trend in history has sometimes been described as an 'Axial Age'.

The concept of an Axial Age dates back to eighteenth- and

nineteenth-century scholarship on religious history.[31] In simple terms, Axiality comprised a novel set of religious and philosophical values which began to appear and spread among the world's largest empires, challenging the hegemony of power holders and articulating a vision for greater social justice. Think of the emphasis on compassion in Buddhism, on the lionization of the meek in Christianity, or the charitable acts that Hindus perform in pursuit of good karma. But because the notion of an 'Axial Age' has been taken up and developed over a long period of time by multiple scholars, each emphasizing different aspects of the phenomenon, it is a difficult transformation to make sense of – and difficult to decide which features to regard as essential.

Here, again, we felt that our Seshat database might be put to use. We set out to determine what effect the Axial Age had on social complexity. The usual Seshat approach to problems of definition is to try to be as inclusive as possible. Recall that instead of picking sides in the debates about how to characterize social complexity, we decided to include in our database details of all the features that scholars had considered to be relevant. We attempted to do the same to capture the core elements of Axiality. Drawing on the literature as a whole it is possible to distil twelve principles (listed here in a slightly simplified form):

1. Moralistic punishment. Moral transgressions should be punished supernaturally or by authorized wielders of supernatural power.
2. Moralizing norms. Religious doctrines and the like that explicitly endorse or forbid various kinds of moral transgression.
3. Promotion of prosocial behaviour and the religious obligation to help others.
4. Moralizing gods who know what you do and are concerned about it.
5. Rulers are not gods. Power holders on earth are not seen as supernatural beings.

6. Elites and commoners are treated the same in the eyes of religion and the law.
7. Rulers and commoners are treated the same in the eyes of religion and the law.
8. Formal legal code. The laws of the land and punishments for breaking them are formalized in a way that can be applied more or less consistently.
9. General applicability of the law. Safeguards exist to limit the ability of power holders or elites to influence the legal process.
10. Constraints on the executive. The presence of official roles and norms intended to limit abuses of power.
11. Full-time bureaucrats. The presence of professional administrators whose job is to implement policies and laws in place of the power holders.
12. Impeachment. The power of the people to remove or punish rulers exercising power arbitrarily.

No one theory of the Axial Age had previously proposed all twelve of these features – in fact, all such theories would arguably exclude at least some of the features listed. However, if we are willing to treat this amalgam as a new and more inclusive theory of Axiality, analysis of Seshat data tells us two rather striking things.[32] First, these features emerged and spread in a wider range of societies than are normally associated with the Axial Age. Second, they first appeared much earlier in some parts of the world than the first millennium BCE, when they are usually assumed to have taken hold. The following diagram shows how the twelve principles (treated as measures that act as proxies for the presence of Axiality) were distributed across ten regions of the world (only five of which were usually considered in writings on the Axial Age) over the course of the last few thousand years. Zero in the diagram represents the start of the Common Era – formerly 'Anno Domini' – corresponding roughly to the date of birth of Jesus of Nazareth.

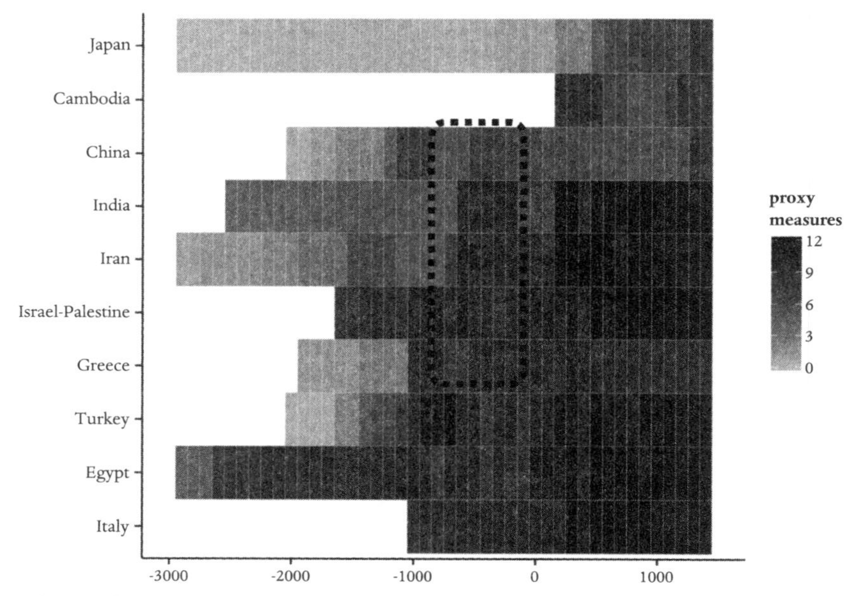

Adapted from Mullins, D. A., Hoyer, D., Collins, C., Currie, T., Feeney, K., François, P., Savage, P. E., Whitehouse, H., & Turchin, P., 'A Systematic Assessment of the Axial Age Thesis Using Global Comparative Historical Evidence', *American Sociological Review*, Vol. 83, No. 3, pp. 596–626 (2018)

In the centre of the diagram, surrounded by a dotted line, are the concentrations of features of Axiality that were present in the first millennium BCE in the regions typically associated with the Axial Age – namely China, India, Iran, Israel-Palestine, and Greece. For every state or empire that was present in the region in question, we used historical evidence to establish which of the twelve principles of Axiality were present. Each of the shaded rectangles represents how many of our features were present in a given region and over what time period. Darker shading indicates more of the features were present while lighter shading indicates fewer of them were present. Note that at least some of the twelve Axial principles were present in our selected regions long before that; in some cases the fullest flowering of Axiality occurred somewhat later than the first millennium BCE. But what is even more striking is the extent to which other regions of the world, not previously associated with the Axiality hypothesis, exhibited the tell-tale features, in some cases much earlier – as in the case of Egypt, Turkey, and Japan, for example.

What drove this shift towards moralizing religions? I have argued that ideologies of extreme deference to rulers were made possible by the homogeneity of early states, whose uniformity was rooted in mechanisms of routinized ritual. And as we have seen, in a state largely composed of people with a shared culture and history, such political systems seemed to work quite well. However, as the Bronze Age came to an end (roughly 3,000 years ago in western Eurasia), these archaic political systems were running out of steam. More and more societies were now being formed through the expansionist ambitions of the world's most land-hungry empires – empires that were far too diverse to be held together by top-down coercion alone.

By definition, an empire is a conglomeration of formerly distinct political systems under a single ruler or government – a process that is often accomplished through conquest. Among the most famous regions of the world in which empires first emerged are the ancient kingdoms of Egypt and the Akkadian, Assyrian, Hittite, Phoenician, and Achaemenid empires of Southwest Asia. In South Asia, similarly formidable military powers include the Mauryan and Gupta empires. In ancient China, the Zhou, Qin, Han, Sui, Tang, and Song dynasties all unified pre-established states through military dominance. The formation of vast and sprawling political systems brought together competing cultural groups, with their own distinctive traditions, customs, languages, and dialects. As a result, rulers needed a novel way to prevent these sprawling coalitions of identities from collapsing into chaos. In this, Axiality arguably proved invaluable.

To understand how Axiality contributed to stable social systems that could hold internally diverse societies together, it is helpful to turn, once again, to China. During the Han dynasty, which prevailed for more than 400 years (202 BCE to 220 CE), Axial principles of Confucianism began to morph into a state-sponsored cosmology, political philosophy, and morality, emphasizing rational thought rather than naked coercion and top-down domination. A similar story may be told later in China's history. Following a turbulent period of internal strife and disorganization during which Axial values took a back seat, Buddhism began to spread more widely in the late sixth century CE, and in 601 CE, Emperor Wen ordered the distribution of relics of the Buddha to temples throughout the land in a deliberate attempt to unify China behind a moralizing system based on karma and the path to enlightenment.[33] While in large multi-ethnic societies there are always many layers in the religious and cultural landscape, the spread of Buddhist ideas arguably contributed to a lasting period of stability during the Tang dynasty (618–907 CE).

To understand why Axiality was so valuable in multi-ethnic empires it helps to return to the seven moral rules we first encountered in Chapter 2. Although the rules themselves are universal – rooted in our evolved intuitions – moral conundrums frequently present us with situations in which more than one principle applies and we have to decide which is the most important. This also means that entire political systems can emphasize certain moral domains more than others; consider, for example, how different the values are of today's most authoritarian regimes compared with the most liberal democracies. Bronze Age states tended to elevate two principles above the rest: deference to authority (especially rulers and elites) and respect for property (especially of the wealthy and powerful). And their ideas about the gods strongly reinforced this. By contrast, Axiality shifted the emphasis much more firmly onto universal principles of reciprocity and fairness often promoted by otherworldly spiritual leaders and ascetics – ideas that would have held appeal across increasingly diverse factions of society.

As a result, we can think of Axiality as a powerful form of social glue holding together very complex and internally divided societies. Rather than being a single bygone 'age' in human history,[34] Axiality is rather a 'stage' through which societies pass once they exceed a population of around one million individuals: a so-called 'megasociety'. Until this threshold, societies can be held together by violent coercion. But after it, top-down domination produces social formations that are too brittle to survive – and elites needed another way to maintain order. Their solution lay in the morals of the gods: beings who encouraged universal values capable of unifying a more internally diverse citizenry, and under which multiple ethnic groups could rub shoulders more comfortably.

If you had to choose between Huītzilōpōchtli (the sun god of the Aztecs who wielded a fire serpent and demanded human hearts to be ripped out on a regular basis) or the Buddha (who advocated kindness towards others and various meditative pathways to self-knowledge and transcendence), which spiritual leader would you prefer to follow? Of course, few people have ever had to make that choice. But the kinds of religions we end up following have had a great impact on the kinds of societies that prevailed in the past and that gave rise to the ones we live in today.

Seshat data is providing increasingly detailed insight into what happened as some of the most oppressive societies in world history ran up against the challenges presented by an increasingly diverse populace, resulting from conquest and absorption.[35] The groups that succeeded in maintaining high levels of cooperation beyond the megasociety threshold in world history were often ones that curtailed political domination and economic exploitation, especially in the most culturally heterogeneous societies. Even though this trend towards more egalitarian political philosophies may not seem to have produced particularly just and equal societies by the light of standards prevalent in today's liberal democracies, moralizing religions were a step in that direction compared with the states that came before. By and large, megasocieties emphasizing fairness and reciprocity in their religious, legal, and bureaucratic machinery were not only better at governing internally and staving off rebellions and revolutions but also better able to mobilize resources to compete with other civilizations, often violently via imperialistic colonial expansion.

As a result, the world became awash with ideas about renouncers and denouncers of earthly authority structures – prophets, gurus, and messiahs spreading visions of peace, social equality, and other-worldly glories that directly challenged power holders and their military ambitions.[36] Moralizing religions eventually became the official belief systems of the most powerful empires that ensued – from the adoption of Taoism in China during the first millennium BCE to the predominantly Muslim Ottomans of the second millennium CE. All the world religions we know today, including the major Abrahamic faiths and the karmic traditions of the East, came to accept discourses of justice, fairness, and reciprocity extolled by the moralizing religions. This meant that the days of widespread human sacrifice and the deification of rulers were over.

But it would be a mistake to think of the Axial revolution as a wholly benign, peaceable force. The unifying religions of multi-ethnic empires also spawned new forms of imperialism. By the fifteenth century, increasingly global colonial projects were being hatched in Europe, starting with the ambitious voyages of Portuguese and Spanish explorers, pirates, colonizers, and slave traders, and achieving its apogee in the British empire. During this phase in the history of colonial imperialism, one moralizing religion – Christianity – played a prominent role. Alongside many undoubted benefits, missionary zeal also served the interests of power holders and extractive industries. So, although moralizing religions introduced a more egalitarian turn in human history, they were not good news for everyone. New ethical codes may have checked the extreme powers of elites and deified rulers and given new hope for social justice among the oppressed, exploited, and downtrodden but for many these hopes were pinned on the afterlife, rather than the here and now. For that reason, social scientists have sometimes described the world religions as 'passivist', encouraging acceptance of the injustices of this world in the expectation of rewards in the next.[37] Karl Marx's famous 'opium of the people' remark was intended to capture this facet of religion: a leap of faith highlighting the need for social justice but providing only an illusory means of obtaining it.

In short, while moralizing religions gave voice to at least some demands for greater fairness and reciprocity in systems of governance, they could not be said to have delivered social justice for many women, indigenous peoples, peasants, indentured labourers, slaves and vassals, and other relatively deprived or oppressed sectors of society. But might these faiths at least have led to a more peaceful world? Unfortunately not. In the next chapter, we will explore the great paradox of the moralizing religions: that many of the bloodiest conflicts in human history have been waged under their banner. The big question is why.

6.

Tribalism and the Evolution of Warfare

A drum was being beaten two rows behind and the crowds around me were chanting deafeningly in unison. I felt goosebumps across my back and arms as the British national anthem was sung. Suddenly, I found myself unthinkingly drawn to my feet as a Mexican wave swept around the Mineirão stadium. I had travelled to Belo Horizonte, Brazil, amid thousands of England football fans at the 2014 FIFA World Cup. I was there for a very different reason from those around me, however. I had come not in hopes of victory but to study the effects of shared defeat when England were knocked out of the competition. Our previous research indicated that losing at football would have a more powerful bonding effect on supporters than winning. As I looked around at the expectant faces of my fellow countrymen, I reflected on how much better off we'd all be once we had not only accepted our defeat, but fully processed the shared misery.

Prior to kick-off, I had been steadily moving through the aisles of seats, interviewing scores of supporters in turn. Most people agreed to let me administer my questionnaire, partly because they were bored waiting for the match to start (although some also clearly found it amusing that an Oxford professor wanted to know how they felt about the endless sufferings of England fans). Occasionally I was intercepted by FIFA officials, suspicious that I was up to no good – but when I showed them my university card and explained the purpose of the study, they were content to let me continue. Once the ball was in play, however, it was another matter. Nobody wanted to talk to me any more and, even if they did, it would have been impossible to hear them.

The scenes I witnessed in Belo Horizonte would be recognizable to millions of football fans all around the world. But what many of those people may not realize is that the displays of mass emotion and excitement, of pounding drums and bodies moving in synchrony, would also have been familiar to our ancestors going back thousands of years, from the mass rituals of ancient Egypt to the state displays of modern China. Many of these vast gatherings in human history would have similarly combined moments of intense euphoria with spells of more negative shared emotions. Examples would have included collective grieving at the loss of great leaders, anger and hatred directed at the group's enemies, or fear and horror at the gory deaths of martyrs, sacrificial victims, and gladiators.

Shared dysphoria in large crowds has long been recognized as a potent source of group bonding but only recently have we begun to understand why. In Chapter 3, I presented a range of evidence that the sharing of emotionally intense experiences among small groups of initiates and soldiers can serve to fuse personal and group identities together. And when people become fused in this way, they become willing to undertake extraordinary acts of self-sacrifice to defend and protect their groups. My focus in that part of the book was on the ways in which fusion worked at a small-group level. But as societies grew, they ran into a problem: how could they create this form of fusion at the scale of much larger groups, including entire civilizations?

It is to this process that we now turn. As societies increased in size, the ability to fuse with the group came to be scaled up and applied to vast populations – not only to small relational networks comprising people who knew each other personally. One of the ways in which this occurred was through participation in mass rituals. That is why I found myself at the FIFA World Cup in Brazil.

The match I had come to attend was bound to be a painful one for my fellow countrymen. Even before the referee blew his whistle to start the game, everyone knew that it was over for England, having already lost their last game with Uruguay. Many fans still hoped for their team to exit the competition with dignity. But the match itself was a ritualized expression of loss – more like a funeral than a march to victory. Their opponents today – Costa Rica – had surprised everyone by beating Italy and sealing England's fate. At one point, supporters of Costa Rica chanted 'Eliminado' at the England fans, triggering scuffles that brought police into the crowds. Yet it was true – England had already been eliminated from the competition and now they were fighting only for their pride. They were denied even that, however, emerging with a nil-nil draw from what may have been one of the most disappointing World Cup football matches in living memory. Many England fans were dejected, some were tearful. From my perspective, it could hardly have gone better.

These kinds of shared experiences of suffering – from the supporters of national football teams today to the multi-ethnic armies of ancient empires – constitute a mechanism that has long been used to bind societies together. In the world of football fandom, group bonding may be largely about fantasy and entertainment, but for our ancestors facing each other on the battlefield it was a matter of life and death – and a force that dramatically shaped the way our societies evolved.

Warfare and the evolution of tribalism

In the last chapter, I argued that as new agglomerations of states grew into the world's first empires, they were often held together in part through the spread of Axial religions. These faiths emphasized moral principles of justice and fairness that transcended other cultural differences and were able to hold together otherwise highly diverse groups. But Axial religions could only get an emerging empire so far. They may have helped the first multi-ethnic empires to grow larger than earlier states with their outmoded forms of top-down oppression; and yet focusing on spiritual enlightenment or kindness towards others was not an obvious formula for militaristic expansionism. In practice, though, Axial religions did become entangled with imperialistic projects and warfare on an unprecedented scale. Why?

The answer lies not in our religiosity bias itself, but in our tribalism. However, establishing the role of tribalism in the evolution of social complexity presents some thorny challenges. Everyone can agree that some kinds of societies grew bigger and lasted longer than others; but it not easy to establish what helped the winners and what brought down the losers. This is where databanks like Seshat are needed. They help us answer such questions by placing the most prominent theories head to head and establishing which ones hold water and which ones don't. Such a process requires too many variables for the human brain to compute unaided.

As I explained in the last chapter, when we were building the Seshat global history databank we decided it was necessary to differentiate fifty-one distinct dimensions of social complexity because different theories have tended to focus on distinct clusters of those factors. To make matters much worse, the number of societies that emerged and collapsed over the past few thousand years of world history are dauntingly vast in number – even our efforts to reduce the diversity to a sample of thirty regions generated a list of hundreds of distinct political systems. As I explained earlier, simply listing examples or plucking out illustrative cases is not the way to go in solving this problem. So, let's look at what a more even-handed approach – based on statistical analysis of all the evidence in one go – can tell us.

The theories we need to test fall into two distinct clusters. The first cluster of theories focus on the *functions* of various organs of society. They argue that social institutions persist because they help all the others to operate smoothly and effectively – this is their 'function'. Functionalist theories take different forms. Some focus on the way top-down systems of governance helped to solve various coordination problems as societies grew larger: for example, how the evolution of more and more levels of hierarchy in Polynesia made it possible for lots of smaller chiefdoms eventually to be ruled over by a paramount chief. Others focus on how systems of gift exchange bound together lots of different groups into a single cultural political entity, whose members should never go to war and should act as allies if any of them were to be attacked. Still others focus on how beliefs about the formation of the foetus function to assign people to different groups, based on descent from common ancestors, with lots of associated rules about how their property should be divided up after they die. Such theories might sound diverse, and they are. But crucially, in every case functionalist theories point to the way a societal innovation – whether that's a method of organizing the government or irrigating the land – serves the specific needs of the society in question. Between them, functionalist theories have been used to explain the rise of societies as diverse as those of the Peruvian Inca, ancient Egypt, and the Moghul empire.[1]

A second cluster of theories has focused on the role of *conflict* in the evolution of larger and more complex societies. Such studies have focused both on internally generated conflict, fuelled by social injustice,[2] and on the impact of inter-state warfare and conquest.[3] Conflict theories generally argue that societies get bigger and richer by managing infighting and class conflict or by killing, pillaging, enslaving, and exploiting vanquished populations. In the course of all this violence, the theory goes, societies suck in more resources and expand to create larger-scale social formations.

The Seshat team set out to test all the above theories in one go.[4] To test the functionalist theories, we included in our database various indicators of technological and economic advancement, for example relating to things like food storage, roads, bridges, canals, ports, water supply, monetary system, and record-keeping. To test theories relating to the role of internal conflict in the evolution of social complexity, we set out to capture information about the kinds of social inequality most likely to generate social tensions: for example, the presence of a class system. And as a rough and ready measure of the importance of external conflict in the society, we used data on military technology – following the logic that societies facing high levels of outgroup threat would invest more in increasing the sophistication of their weaponry. We also included a measure of mounted warfare because much previous research suggested that horse-riding combined with metal weapons (e.g., armour and projectiles) was associated with increased warfare intensity.

The advantage of analysing the relationships between all these variables statistically on a computer was that it made it possible to see patterns that brain power alone could not possibly recognize and tease apart. It would be difficult enough for us to track mentally the relationships between just two variables over multiple polities during long time periods. A computer, on the other hand, was able to look at seventeen potential drivers of socio-political complexity in hundreds of political systems over thousands of years. Armed with data on our long list of variables covering 373 societies, Peter Turchin undertook a sweeping statistical analysis of how each combination of variables correlated with levels of social complexity.

We were not surprised to find that agriculture turned out to be prominent among the statistical drivers of societal size and complexity – after all, the scale and efficiency of food production inevitably set limits on how many mouths could be fed as well as the extent to which the lavish lifestyles of non-productive elites could be supported. But what came as a shock – to me at least – was the other force that was consistently behind the growth in the size and complexity of ancient civilizations: military technology.

The largest societies to emerge over the course of world history all tended to follow a series of major growth spurts, each associated with new forms of military technology. The first big spurt was associated with the spread of bronze weaponry, for example in Mesopotamia and Egypt, from around 3000 BCE. The next big spurt came with the arrival of chariots, providing a mobile platform from which to shoot arrows from increasingly powerful composite bows. In the second millennium BCE, this leap forward catapulted the Shang empire in China, the Hittites in Anatolia, and New Kingdom Egypt to scales of organization and territorial spread unprecedented anywhere else in world history. Similar leaps forward were later associated with various Eurasian subregions in which more advanced forms of armour and mounted warfare became established. And then another spurt occurred following the discovery and effective deployment of gunpowder. All these innovations in military technology produced more rapid effects on social complexity than anything previously accomplished as a result of agricultural intensification alone. It wasn't only in Afro-Eurasian empires that military technology had this striking impact on upsweeps in the scale of political organization. We found parallel patterns in Mexico where the introduction of bows and arrows and stone-bladed broadswords also preceded sharp rises in social complexity. Unlike the largest empires on the other side of the Atlantic, however, the Aztecs lacked the capacity to exploit pack animals to move troops and supplies. The only empire in the Americas that achieved greater scale, prior to the introduction of the horse by Spanish conquistadors, was the Inca empire, which exploited indigenous llamas for transportation.

But what was most striking of all was that *none* of the other variables we had chosen as proxies (apart from agriculture) seem to have had any significant effect on the evolution of socio-political complexity. As such, most of the functionalist theories and those focused on internal conflict were not supported. This is one of the major contributions of our quantitative approach to world history, which allows us to winnow out theories that don't work and to focus our attention increasingly on the more promising ones. We had found the first rigorous, quantitative evidence that violent military conflict has been a primary driver in the growth of human society – much more than any other force, bar the spread of agriculture.

It was disconcerting to appreciate the extent to which something as toxic as intergroup violence could have had such a positive effect on the evolution of human civilizations. The notion that the many forms of creative expression we naturally admire – from artistic accomplishments to exquisite works of architecture – might be underpinned by violent conflict is alarming. But there is a compelling rationale for it. In his book *Ultrasociety*, Turchin argues that the reason warfare enlarged the scale of warlike tribes in world history is because – all else being equal – big groups win out in competition with small groups.[5] In fact this logic is apparent in many theories of early state formation. Using examples of early Mesopotamian, Classic Mesoamerican, Mature Harappan, and Predynastic Egyptian civilizations, it has been argued that the process of converting localized chiefdoms into more organized and territorially extensive states was motivated by the desire of urban elites to gain access to more distant resources. Inevitably, this meant colonizing more lands and creating systems to funnel exotic artefacts and produce through peripheral outposts.[6]

However, taking possession of more distant territories, holding onto them, and maintaining the routes for wealth to flow back to the centre also required military innovation. These new outposts needed to be defended. So, armies became bigger and rulers' violent appetites more rapacious. States expanded to become empires – incorporating vanquished peoples into the overall population. In turn, these empires grew richer still through the spoils of war, and this further boosted demographic growth and funded ever more sophisticated organs of government and economic management – from the creation of vast irrigation complexes and monetary systems to the creation of professionalized bureaucracies and institutions of learning. Violent conflict drove growth, and growth drove violent conflict: an endless snowball of social complexity, all fuelled by military expansionism.

All this only raised further questions, however. To my mind, the biggest enigma was the behaviour of the soldiers themselves. Our findings made it unambiguously clear how violent conflict helped elites: after all, they were the ones acquiring most of the spoils of their conquests. But what about the troops themselves? What was motivating these first standing armies – in many cases, made up of thousands of troops – to fight and die for some distant ruler?

Any persuasive answer, I suspected, would need to relate to the *group identities* of individual soldiers. Armies need to be willing to risk life and limb, and that requires commitment to the group. Motivating soldiers to fight back when they come under siege or to stand firm in the face of the many ferocious smaller and highly cohesive groups that formed on the hinterlands of an empire would have required methods of generating feelings of loyalty that transcended internal divisions. To elicit these emotions, the first armies would come to draw upon the strongest form of tribalism known to humankind – identity fusion. But they would do so on a scale quite unlike anything that had come before.

Two forms of group bonding

Anyone who wants to create an effective large-scale fighting force faces a problem. To motivate people to fight and die for the cause, you would ideally want their identities to be fused. Fusion is the most powerful social glue known to humanity, capable of driving extreme forms of self-sacrifice for the group. The trouble is that fusion evolved as a mechanism for bonding in small warrior bands – not in huge modern armies. The question faced by the military leaders of the world's first states and empires was therefore whether fusion could be scaled up.

As I argued in Chapter 3, in many small-scale warlike societies – such as the Baining villages where I undertook my first fieldwork in Papua New Guinea – willingness to fight and die for the tribe was traditionally fostered in adolescence by painful initiation rites. In ancient tribes, as in many small-scale military units we observe today, fusion would often have been rooted in intense memories of shared suffering: what I call the 'imagistic pathway' to identity fusion. There is a hard limit to the size of tribes formed in this fashion, however. You couldn't add people to those memories if they weren't present during your initiation. And you couldn't exclude anyone from those memories, pretending they weren't there, if you could perfectly well remember them going through those ordeals alongside you. This pathway to fusion therefore creates quite small and rigid social groupings – rooted in relationships with people you actually know personally and with whom you can recall having been through life-changing episodes together.

At the same time, it is obvious that strong forms of group bonding *do* exist in much larger imagined communities – something I had witnessed during my fieldwork with the Baining people. Thankfully, the Kivung movement was a peaceful organization, but its members devoted huge amounts of time and resources to supporting the movement, to winning the approval of their ancestors, and to defending the group against its enemies through the ballot box rather than by means of bullets and bombs. Kivung members were not warlike, but they certainly had a strong sense of unity as a political force to be reckoned with. This is precisely why movements like the Kivung were widely interpreted as the first stirrings of militant nationalism in many regions of Melanesia and often suppressed on those grounds by colonial administrations.[7]

Drawing on my field observations and also the insights of another anthropologist, Fredrik Barth,[8] I came to the view that there were two distinct forms of group alignment, rooted in contrasting ways of remembering and transmitting collective rituals. On the one hand, there were the initiation cults that Barth and I had each been studying, which were constructed around rarely performed, emotionally intense experiences remembered as life-changing events. On the other, there were the much more frequently performed and highly routinized activities seen in movements like the Kivung. As we saw in Chapter 4, this resulted in standardized schemas and scripts that could be spread more efficiently by guru-like leaders to much greater populations.

Later, as I became more immersed in psychological methods and findings, I realized that there were special terms for these two ways of remembering rituals. The experiences of initiates in the traditional male cult were recalled as distinct, highly memorable events in their lives, and thus stored in 'episodic memory'. This is the type of memory we draw upon when we 'relive' a specific event: your memories of the time you were initiated into the tribe, got married or graduated from university. By contrast, the teachings and practices of the Kivung were not tied to particular events. Over time, people's experiences of participation melded into each other so that the way people mentally represented daily rituals, such as laying out offerings in the cemetery temple, was more like a generic template for the behaviour. That is, instead of remembering a particular episode of laying out offerings, one only had to recall the elements that generally occur, such as cleaning and polishing plates and cutlery, decorating tables with flowers, and carrying out various other ritualistic duties and observances. Psychologists refer to this as 'semantic memory' and it comprises our general knowledge about the world, such as how to behave on a bus or the fact that Paris is the capital city of France.[9]

It seemed to me that when these two ways of remembering are applied to shared experiences – the stuff that defines the groups we belong to – it gives rise to two very different ways of thinking about collective identity. The small-group identity of initiated warriors was based on specific, episodic memories of the rituals they had undergone together. This is what group psychologists now call identity fusion, as described in Chapter 3. However, their big-group identity as Kivung members was based mainly on semantic memories about more generic features of everyday life, such as regular observances and sermons. It struck me that this way of remembering produced 'identification' – a form of group alignment that has spawned a huge literature in social psychology.[10]

I felt that the literature on 'identification' was missing one key point: the fact that this form of group identity was rooted in semantic memory. It struck me that when we are identifying with our 'ingroup', we are not reflecting upon specific episodic memories but on the sharing of generic identity markers stored in semantic memory. For example, think about the country you belong to and the things that mark out that identity: a flag, a national anthem, maybe a special form of national traditional dress or a type of religion or ethnicity. These are all ways of thinking about identity which are not necessarily anchored in any distinct episodes in our personal life histories, and we don't normally remember when we first realized that our country has its unique flag or anthem.[11]

When we identify strongly with a group, we soon become prone to favouritism: we look out for members of our own 'tribe' and adopt a competitive or even hostile attitude towards people who are not in that group.[12] Imagine, for example, two sisters who become passionate about quite different teams. One becomes a committed supporter of Arsenal FC and the other of Tottenham Hotspur. Whenever the two sisters talk about football, the conversation tends to descend into unwinnable arguments about which team is better. They end up agreeing to disagree. However, when the topic of football doesn't come up, they seldom get into an argument. It is as if they get along fine when the focus is on their respective personal lives and only fall out when group identities become salient.

In this respect, identification is fundamentally different from fusion. As we've seen, identity fusion involves sharing something self-defining with a group: whether that is an intense memory (like an initiation rite) or a biological bond (like brotherhood). As such, identity fusion is closely bound up with our deepest sense of *autobiographical self*. On the other hand, identification has very little to do with our sense of ourselves as unique persons. For these identities to increase in salience, it doesn't really help to remember personally life-changing experiences – we need only be reminded of our identity by more generic symbols like flags, scarves or team T-shirts.

In fact, identification usually need not activate our personal identity at all.[13] Quite the opposite, in fact. If you are strongly identified with a group such as your country or football club, the more you think about national events and league matches, the less accessible your sense of self becomes. For example, if I happen to be an Arsenal fan, then the more you make me aware of my identity as a supporter by reminding me that you support Tottenham Hotspur, the less I'll be likely to assume the persona of Harvey – a generally tolerant chap. Instead, you will find yourself talking to just another member of a particular tribe who despises rival tribes and is quick to recall negative stereotypes about them.[14]

Psychologists describe this as 'deindividuation' – a circumstance in which our personal identities and motivations can become inaccessible, and we feel compelled to act in conformity with social cues.[15] For example, a highly identified person is likely to fall in line and follow orders, because that's what a prototypical group member should do.[16] And it does not take much to induce this kind of deindividuation. Simply putting on a uniform is enough to make many people administer more prolonged electric shocks to people who aren't wearing the same uniform. This is particularly likely if the clothing is associated with violence-condoning norms. For example, dressing up in a Ku Klux Klan outfit results in more aggressive use of electric shocks than dressing up in a nurse's uniform.[17]

Identification is especially important in large-scale, complex societies where everyone frequently interacts with strangers – for example, when haggling in a bustling urban market, when remonstrating with state officials, or when going to a barber for a haircut. In many such interactions, we may assume quite two-dimensional identities, such as customer, citizen, client, and so on. In extreme cases, it can seem as if our social identities completely swamp our sense of individuality. We simply disappear into the crowd.

All these factors – deindividuation, outgroup derogation, and potential for increased outgroup hostility – are potential assets for military groups, contributing to the will to fight. Identification makes us want to win out against members of other groups. The trouble, however, is that identification can only get you so far. It can certainly make people willing to derogate an outgroup and favour an ingroup. But it seems not to make people willing to put their lives on the line for the ingroup.[18] Since identification doesn't tap into any intense personal memories or our deep sense of self, its group-oriented demands can be overruled by more selfish considerations. From a personal perspective, the strongest imperative imaginable is one's own survival – even if that conflicts with what the group would like us to do. For many, a very natural response in any dangerous situation would be to try to hide in the crowd or to run away when the opportunity arises.

Based on this reasoning, identification would be inadequate to motivate acts of violent hostility against an outgroup when this would entail risk to life and limb. Evidence for this comes from studies in which the effects of fusion and identification on willingness to sacrifice for the group have been directly compared.[19] To give just one example, my colleagues and I ran a survey with over 200 Muslims living in the UK and Belgium, asking them about experiences of abuse as members of a religious minority. In particular, we found identification to be a much weaker force than fusion in willingness to lay down one's life to defend fellow Muslims.[20]

Nevertheless, this study with Muslim minorities only asked about hypothetical willingness to engage in self-sacrificial acts for the group. In other studies, we have compared the effects of identification and fusion on willingness to place oneself in harm's way in reality, rather than just in theory. For example, in one study with around a hundred Australian football fans from two rival clubs in Sydney, we set out to compare the role of identification and fusion respectively in driving hostility to an outgroup. The sample included members of opposing fan groups that had been criticized in the media for throwing flares on the beach, ripping out seating in the stadium, and fighting with rival fans. Our study found that identification with one's football club was associated with outgroup *prejudice*: that is, our subjects were likely to tell us they disliked members of the opposing club. However, fusion was more likely than identification to be associated with actual violence against rival fans – behaviour which of course implies higher levels of personal risk than merely holding prejudiced attitudes.[21]

Identity fusion is therefore undoubtedly useful if you want to build a successful military. If you are fused with an army, then its battle plan – no matter how dangerous and daunting – becomes your own personal objective. The trouble was, as armies grew larger, they also became more anonymous. The group's bonds became embedded in symbols of shared identity rather than in directly shared experiences. And as such, small-group fusion was at risk of being eclipsed by big-group identification.

And yet, somehow, the world's armies have for millennia found a way to overcome this problem. Starting with the first ancient militaries defending themselves from violent neighbours, armies have managed to convince vast groups of people to die for their cause. How? The answer was a form of fusion that can operate on the same scale as identification. We call it 'extended fusion' and it constitutes one of the most dangerous inventions complex societies have ever come up with.

Extended fusion and warfare

The revolutionaries I met in Libya in 2011 – the year of the Arab Spring – were fused with the other men in their battalion. These included the people they knew personally and had suffered with in deadly combat or when sheltering from the shelling of tanks and the bombs that rained down on their city. No surprises there; these were precisely the kinds of intense personal experiences that we had long known lead to identity fusion. But these men also indicated they were highly fused with the members of other battalions they didn't know personally – utter strangers who had endured similar ordeals but not ones that were shared in person. This was the essence of 'extended fusion': identity fusion between large groups of people who had never personally met. But how could extended fusion come about, in the absence of shared, in-person experiences?

Initially, we thought that the key to extended fusion lay in the projection of local ties (like your immediate fighting unit) onto extended categories (like your entire battalion). That is, we imagined that in order to be fused with a very large group, you needed to draw on feelings of fusion with a much smaller group of people you knew personally like your family – and then 'project' it onto some wider community. We have found some evidence for this hypothesis. For example, in 2013 – a year after we published our first efforts to theorize the relationship between local and extended fusion in terms of a process of 'projection' – an atrocity occurred in Boston, at which runners in an annual marathon were targeted by terrorists who planted two home-made pressure-cooker bombs close to the finish line. Hundreds of people suffered injuries, seventeen lost limbs, and three died of their wounds. By chance, one of our team members – Michael Buhrmester – had just begun collecting data on extended fusion with fellow countrymen using a sample of participants from Boston, just days before the attack. As soon as news broke of the bombing, Buhrmester suggested repivoting the study to see whether those previously more fused with their fellow Americans were also readier to put their money where their mouths were (so to speak) by being more generous in their efforts to support the victims.[22] What we found was that Bostonians who expressed a strong sense of psychological kinship with the victims – that is, those who felt that their fellow Americans were like family – scored higher on levels of 'identity fusion' and were much more likely to donate to the relief effort. We argued that this response was rooted in the biological pathway to fusion. Pre-existing feelings of fusion with actual family were being extended to a much larger 'family', comprising the citizenry of the entire country.

However, in time we realized that this was not the only route to extended fusion. As with local fusion, extended fusion is often created by shared experiences – just ones that are experienced at the scale of large, anonymous groups. We now have multiple studies showing that the sharing of transformative experiences can increase fusion among vast, anonymous communities of people who have been affected.[23] It is now clear that extended fusion – whether it is based on perceptions of shared biology or shared experience on a large scale – is one of the most powerful forms of tribalism in human history. I view it as analogous to the discovery of gunpowder or powered flight: innocuous-seeming innovations, but ones that led not only to dazzling firework displays and high-speed travel but also to highly efficient killing machines. At its most benign, extended fusion motivates heart-warming events such as World Cup football matches and Live Aid concerts. But the dark side is the role this new form of tribalism has played in the most destructive wars of human history.

The first signs of the growing importance of extended fusion in the history of warfare are to be seen in the armies of the ancient world. For example, there is evidence that in ancient Greece the most effective armies derived their courage and resolve from feelings of psychological kinship on the battlefield. Back in the fourth century BCE, Aeneas Tacticus wrote that the strongest armies are those who fight 'for all that is most dear to them'. When soldiers were fighting 'for temples and fatherland, for parents and children and all they possess,' he wrote, 'a successful and stout resistance to the enemy will make them dreaded by their foes and more secure from future invasion.'[24] In other words, the most formidable fighting forces are highly cohesive, spurred on by a stubborn refusal to admit defeat.[25] From the viewpoint of our theory of extended fusion, what is of particular interest is his emphasis on the idea that effective soldiers are motivated by the desire to protect not only their families but also their country and religion, which is writ large as a kind of vast web of kinship: a 'fatherland'.

Shared experiences of battle also likely fused combatants to their comrades in ancient city states, particularly with the rise of siege warfare. Aeneas Tacticus' appositely entitled *How to Survive Siege Warfare* – the world's earliest known military treatise – argued that the key to military success was a strong sense of cohesion and shared purpose. Tacticus lived in a world of numerous city states (it is estimated that over a thousand of them flourished in ancient Greece). Although some came to achieve dominance over others (think Athens, Sparta, Corinth, Thebes, Syracuse, or Rhodes), people's primary allegiances tended to lie with their own urban settlements rather than with larger agglomerations of them. Nevertheless, these ancient cities were still too large for all their members to know each other personally. And for their armies to be willing to fight and die in their defence would have required very strong forms of cohesion to be extended to anonymous others.

The role of extended fusion in specific battles and wars is also detectable in the writings of the ancient Greek historian Polybius, who argued that combat can dampen the selfish and egoistic tendencies of any one individual, and instead motivate commitment to an extended group. Polybius argued that the reason so many armies arrayed against the Romans were defeated was because they deployed mercenaries – and therefore lacked the cohesion of militias defending their own families and territories.[26] Conversely, one can see traces of how ancient military strategists benefited from the effects of extended fusion in descriptions of the way battles were conducted between Spartan armies and Athenian hoplite militias in ancient Greece. There is little doubt that Spartan forces were better trained, organized, and led than their Athenian adversaries. However, what the Athenians lacked in terms of training and chain of command, they made up for in terms of group cohesion – including bonds resulting from shared experiences of past battles. This was noted by Athenian commentators themselves – such as the veteran commander Xenophon – who tried, in effect, to persuade his fellow countrymen of the need to harness the power of extended fusion, by combining it with the Spartan model.[27] Polybius had direct experience of prolonged urban warfare, including the famous siege of Carthage,[28] and made appeal specifically to shared suffering as a basis for military cohesion. All these historical accounts and many more would indicate that extended fusion within ancient militaries was underpinned by a sense of both shared biology and vicariously shared experience.

Sacred values and the rise of imperialism

When people fuse with a group, their personal group identities become functionally equivalent – if you attack the person, you unleash the rage of the group and if you attack the group, it is taken personally. This is why fused armies present such a formidable fighting force. And so, it is easy to see how extended fusion motivated the first large armies to march into battle against all odds, when the state or empire came under attack. Fusion, whether local or extended, comes into its own when armed groups have their backs to the wall.

Fusion activates feelings of psychological kinship, as we have seen. In times of peace, people who love their families do not wish to annihilate other families – they would far rather live and let live. Fusion likewise only triggers extreme self-sacrifice when the beloved group or its members come under threat. I was repeatedly told by the armed revolutionaries I talked with in Libya that they first took up arms when they heard that Gaddafi's forces were raping and murdering civilian women and realized that their wives, mothers, or sisters could be next. In their eyes, the insurgency was fundamentally an act of defence. This is also a common refrain when listening to the reasons why anyone volunteers to go into battle – whether we describe them as freedom fighters, tribal warriors, conventional armies, or terrorist cells.

But there also seems to be another source of willingness to wage war which appears on the face of it essentially offensive and acquisitive, and thus hard to explain as an outcome of fusion, whether local or extended. Specifically, how do we account for the widespread tendency of imperialistic empires to engage in acts of unprovoked predation and conquest? Scores of such empires could be listed spanning time periods ancient and modern, starting with the great empires of Eurasia through to those of sub-Saharan Africa and Mesoamerica in the pre-modern period, and of Europe, Russia, America, and Japan in more modern times. If fusion is essentially motivated by defensive rather than offensive thinking, imperialism is not explainable in terms of fusion alone. So, what is the psychology behind it?

An early attempt to answer that question may be found in the works of fourteenth-century Arab historian Ibn Khaldun, who set out to explain the evolution of tribalism and warfare over the ages.[29] He recounted many of the long-term historical processes I have also been describing: starting with local bonds being forged within tribal groups, leading onto shared, larger-scale identities at the level of city states, and culminating in tightly bonded militaries at the level of much larger states and empires. In particular, Khaldun emphasized the importance of emotionally intense rituals in the creation of *asabiyyah* ('solidarity' in Arabic) – and particularly the role of military music such as the beating of drums and the sound of trumpets and horns, and the use of visual symbols such as flags and banners. Although Khaldun did not explicitly describe *asabiyyah* as the fusion of personal and group identities, he clearly thought, like me, that social cohesion could be generated through intense shared experiences.

But he added to this another dimension that is peculiar to imperial power – one that is crucial for understanding the logic of military conquest as we know it today. Khaldun argued that there was a moral imperative behind imperial conquest. He wrote that to succeed, empires needed to unite conquered tribes and states into an overarching, centralized political structure – one bound together by higher values and capable of extending *asabiyyah* to the empire at large. Not only was this necessary, Khaldun argued, but it was also virtuous. Having established these civilizing virtues, military expansion should become a 'universal mission'. In other words, killing the heathens and unbelievers became a sacred duty that went hand in glove with the imperial imperative. Khaldun illustrated this point by approvingly describing the story of a Muslim military commander who, upon discovering an Indian community that engaged in practices forbidden under Muslim law, carried out a mass slaughter and placed the heads of local leaders on public display.

The missionary zeal with which the great empires of history have sought to impose their laws and values in the lands they invade was a decisive development in world history and one that requires a special explanation. The idea that there is only one privileged way of viewing the world – one that is universally true and morally superior to all others – only emerged with the spread of the imperial empires. It is also closely related to the rise and spread of moralizing religions. In most ancient city states, in regions such as Mesopotamia or Crete, local cohesion was strongly linked to local gods and the temples unique to the cities in which they were found. By contrast, moralizing religions insisted on principles applicable to everyone, irrespective of what local religion they adhered to. Consequently, all parochial beliefs that directly contradicted those more general ones came to be seen as immoral and subversive – often justifying the silencing or elimination of those espousing them.

The last chapter ended with a question: how was it that the universalist and moralizing religions which first began to emerge during the so-called Axial Age eventually came to be invoked to justify violent conquest? Ibn Khaldun's writings offer the first hint of an answer: that the sense of moral purpose at the heart of these religions can be used to justify the expanding dominions of empire. In this context, conquest becomes not merely an acquisitive act or an expression of dominance, but a sacred duty. This is how imperial expansion became not merely a means to some material end, such as to plunder the resources of weaker groups, but a moral quest, often carried out with appeal to divine sanction.

Imperialist conquests aimed at spreading the truth and the light while obliterating unbelievers are commonly described as 'Holy Wars'. Some of the earliest examples of this way of waging warfare come from the ancient Israelites in the first millennium BCE – the same period of world history typically associated with the so-called Axial Age and the spread of moralizing religions discussed in the last chapter.[30] Around the same time in China, the Han dynasty began to integrate Confucianism into imperialistic ambitions. Efforts to establish theological hegemony based on moralizing values had its ups and downs, as we have seen. Nevertheless, efforts to establish the supremacy of sacred values eventually became the hallmark of warfare in many of the world's largest empires.

The concept of a Holy War is perhaps most famously associated with the Christian crusades and with the Muslim concept of jihad. However, the concept of a just war is present also in Hinduism and Sikhism and examples of wars waged under the banner of religion are easy to list in innumerable countries around the world. Yet it would be wrong to see this as an intrinsically religious phenomenon.

In fact, the idea of a Holy War has much more to do with tribalism than with faith (which is why I am focusing on it here, rather than in the previous chapter on religious belief). Indeed, the twentieth century was arguably dominated by a form of secular Holy War between communism and capitalism – two ostensibly non-religious worldviews coalescing around opposing values that each side regarded as sacred, just as crusaders, jihadists, and other religious zealots had done for centuries before. In all such cases, a sense of 'moral crusade' seems to play a role. But where does this sense come from – and what effect does it have on intergroup violence?

The answer to this question may lie in the theory of so-called 'sacred values'. Sacred values are inviolable principles – ranging from abstract ideas like a belief in transubstantiation to an object that must never be mistreated such as a flag or a holy book. The defining feature of sacred values is that they trump all material, monetary, or practical considerations and cannot be compromised under any circumstances. Anthropologist Scott Atran argues that because sacred values are inviolable, they can often become a sticking point when trying to resolve or arbitrate in intergroup conflicts.[31] Atran and his colleagues have produced compelling evidence that efforts to negotiate settlements between groups in conflict by offering material incentives to compromise on sacred values can backfire badly. Instead of incentivizing peaceful agreements, they trigger moral indignation, inflaming intergroup hatreds, and spiralling into worse forms of violence than ever before. These are not beliefs that can be bought and sold.

Just as highly fused individuals will jump on a grenade to save their comrades, so people often appear willing to fight and die to defend their sacred values. Atran argues that this is often underestimated by today's liberal democracies.[32] For example, US foreign policy failed to anticipate the will to fight among militarily weaker adversaries in Vietnam, Iran, Iraq, and Afghanistan. The military logic appears to be that by instilling 'fear and awe' through an initial display of power, it will become clear that resistance is useless, and morale will be broken. And yet time and again, this is not what happens. Atran points out that the same mistake was made by Napoleon, Hitler, and other leaders of empires in the past. What often happens in such cases is that the numerically weaker group returns to the fight after initial heavy losses with a redoubled determination to battle on.

When I returned from my visit to Libya in 2011, I began to wonder if Atran's ideas could help explain the origins of imperialism. It seemed to me that identity fusion and sacred values were closely related or overlapping constructs. What if people's sense of deep commitment to a group – in the form of extended fusion – became somehow attached to certain values, in effect 'sacralizing' them to the point that they would be worth dying for?

This was admittedly a rather complex thought process and requires some unpacking. Fusion, as I had encountered it, involved a commitment to other people who shared life-changing experiences with you – whether people you directly knew in the case of local fusion, or some wider community in the case of extended fusion. By contrast, I had come to think of values and beliefs as a basis for identification rather than fusion – that is, a form of group identity that did not involve undergoing personally transformative events. But what if certain values themselves had somehow become strongly associated with the groups people were fused with? And what if those associations became so strong that the beloved group and its sacred values no longer seemed distinguishable? I first recall reflecting on this idea as I stopped to gaze upon the Martyrs' Memorial in the heart of Oxford on my way home from a college dinner. It was a familiar enough sight, but I found myself staring at it harder than usual. The edifice celebrates the heroism of Radley, Cranmer, and Latimer, who preferred to be burned at the stake in 1555 rather than renounce the Protestant faith. For them, a sacred value was at stake – one that was worth dying for. But was that because the Protestant faith and the group with which they were fused had become essentially the same thing?

To explore these issues in more depth, I invited the original creators of the identity fusion construct – William B. Swann and Ángel

Gómez – to meet with me and Scott Atran in Oxford. The discussions were by no means conclusive, and it is more than likely we all emerged with different conceptions of relationships between sacred values, fusion, and identification. However, my own working understanding of how these constructs operate together is as follows. Fusion, based on the sharing of self-defining experiences, motivates us to defend our groups but only when they come under attack. Identification, which is based on shared identity markers, motivates us to act competitively towards outgroups whether or not they are being attacked. Sacred values are a combination of both these processes. To be sure, they are shared identity markers – values in this case that declare to the world what group we belong to. Protestant values would be a good example – a cluster of beliefs let's say about salvation being attainable through faith in Jesus, forsaking many other ideas associated with the Roman Catholic Church. What would make Protestant values sacred, however, would be if they were somehow to become personally self-defining – for example through some kind of revelation, epiphany, or conversion, or perhaps through a shared history of persecution in which beloved members of the group have suffered because of their Protestant values. In this way, we would become fused to those values – elevating them to sacred status and making them worth dying for.

If this idea is correct – and sacred values result from a combination of fusion with the group and identification with a set of values which act as group markers – then it would mean we get a sort of double whammy; one that combines ingroup love and outgroup derogation to form a potent cocktail. On the one hand, our personal self becomes deeply invested in the extended group – such as our country or empire – and that's the fusion element. On the other hand, the emblems, ideologies, and other distinctive identity markers and values associated with that group – the identification element – comes to be seen as integral to the extended group we are fused with. So now this means that we are not merely prepared to risk life and limb to protect the group, but we are also willing to do anything we can – using violence if necessary – to advance the ideology of the group. This sacralization of values seems to me what lies at the heart of imperialism and violent extremism.

Atran and his colleagues have shown that commitment to sacred values can explain 'holy' acts of violence based on research conducted in some of the world's most volatile regions, from Mosul in Iraq after the expulsion of ISIS to urban Moroccan neighbourhoods associated with previous terrorist bombing campaigns. In all the populations sampled, they found that willingness to fight and die for a group was associated with a quality they describe as 'spiritual formidability', defined as 'the conviction and nonmaterial resources (values, strength of beliefs, and character) of a person or a group to fight and achieve their goals in conflict'.[33] And they found that this spiritual formidability, when combined with extended fusion, was a remarkably good predictor of violent conduct.

Sacred values lead us to view violence against an outgroup as not only an act of self-defence but as a moral obligation to cleanse the earth of heretics and infidels. In many imperialist groups through history, the shared symbols of empire have come to be seen in absolutist terms – as general and inviolable truths rather than merely parochial identity markers. From Christian crusaders to Muslim jihadis, and from the Nazi occupation of Europe to the Russian invasion of Ukraine, imperialism seems to tap into the ancient power of fusion but also the process of identification with doctrinal systems and ideologies. As a result, it sacralizes the group's most cherished values – and this is the key force driving various forms of predatory violence from suicide terrorism and crusades to pogroms and genocides.

The aim of Part Two of this book has been to describe how three ancient biases (conformism, religiosity, and tribalism) were moulded and harnessed as larger and more complex social systems evolved. Some of these societies found the sweet spot between innovation and tradition by exploiting both ritual and instrumental stances to the full. They also created institutions and philosophies that encouraged a more future-minded outlook. They developed doctrinal religions, unrecognizable from the 'wild' religions of our ancestors, fostering new forms of top-down governance and peer-to-peer enforcement. And violent tribal instincts were increasingly channelled into the creation of multi-ethnic empires and vast trading networks.

All these innovations in world history have added to our collective inheritance, culminating in the geopolitical landscapes we inhabit today. Throughout the many great conflicts of human history – from the wars waged by ancient Egyptians and Persians through to the Mongol invasions and the sacking of Rome – armies were driven not only by the greed and ambition of power-hungry leaders but by extended fusion with those who suffered alongside them. As the mightiest of those empires adopted world religions as part of their toolkit for extended fusion, they also introduced new sacred values that could serve as a rallying point for people of diverse heritage. The world religions and their direct ancestors gave warlords and emperors with imperialistic ambitions a new way of uniting and motivating people from different ethnic groups, classes, and territories. This has come to be the hallmark of wars in modern times – including the worst of all world wars in the twentieth century. Seen in this light, the rise of ideological extremisms of recent memory (including the many varieties of fascism, revolutionary communism, militant nationalism, and jihadism) are simply a natural continuation of this trajectory towards extended fusion, imperialism, and violent self-sacrifice.

Spartan phalanxes, Mongol hordes, Samurai swordsmen, Maori warriors, Gurkha soldiers, kamikaze pilots, and any number of warrior tribes in history have refused to surrender even when faced with insurmountable obstacles. Acts of violent self-sacrifice have been documented throughout history. However, the transnational religions and ideologies of the contemporary world have managed to create a volatile mixture of extended fusion and sacred values – with their adherents not merely content to die defending the group but also to defend its most inviolable and non-negotiable dogmas. As such, we now stand on the brink of a new crossroads in world history. How we might seek to navigate these complex but also potentially deadly aspects of human psychology in the modern world will be the subject of the chapters that follow.

PART THREE

Nature Reimagined

7.

Conformism and the Climate

Summer 2011. It was the digging season at Çatalhöyük and I was chatting with colleagues during a tea break when suddenly we heard a commotion outside. Someone was shouting 'Fire!' Turning to the window, we could see people running up to the crest of the east mound. We hurried to join them. Arriving breathless at the excavations, we saw the sky darkening with smoke. A wildfire was tearing across the landscape, fanned by the wind. Everyone around me was focused on one question: was the fire heading this way? Would 10,000 years of priceless archaeology be destroyed in a single conflagration?

Some said the fire was probably started deliberately by farmers hoping to save money clearing their land. If so, they would not admit it but claim that it started from natural causes. In fact, the land around Çatalhöyük had become a tinder box, capable of catching alight from the slightest spark: a carelessly discarded cigarette butt or a bolt of lightning. The truth was that the Konya plain – previously one of the world's most fertile cradles of civilization – was turning to desert. A few years after the fire, following the Paris Agreement on climate change in 2015, the Turkish government agreed to address the problem of desertification in this part of Anatolia, but few of the archaeologists I spoke to in the region were optimistic about the direction of travel.

As I stood on the hillside watching the dry, yellow fields glow red in the spreading flames and the plumes of black smoke spiral into the azure sky, I was reminded how much we still have in common with the Neolithic people who first settled here at Çatalhöyük. They, like us, were eager conformists, whose agricultural practices became suffused with unquestioned norms and rituals, coalescing into traditions that shaped the way they exploited the landscape. And yet these very same tendencies were now driving us to ruin.

According to an Emissions Gap Report issued by the UN in 2022, the world must cut 45 per cent of greenhouse gas emissions by 2030 to avoid irreversible environmental catastrophe.[1] Unfortunately, the opposite is more likely – that we will *increase* and not decrease our global emissions. Why? Because the changes needed to decarbonize are politically unpopular and divisive. Economists and climate experts assure us we are all going to suffer as a result. It is estimated that by 2050 the world will need 50 per cent more food to survive due to projected population growth but the impact of climate change on agriculture is expected to reduce rather than increase yields.[2]

Of course, on one level we all know why this is the case. Politicians are elected for short periods of time and need to throw money at parochial problems to accomplish short-term outcomes if they are to stay in power. Our norms in business and commerce likewise prioritize quick profits over long-term benefits. The climate crisis is a global problem, but we don't have a global government. Rather we have a couple of hundred countries, each exercising sovereignty and prioritizing domestic interests.

On another level, however, we seem to have no idea. Looking at climate change through my anthropological lens, it strikes me that at the heart of the climate crisis are the ancient biases of conformism, religiosity, and tribalism. The very natural urges that once built civilization now pose its most dangerous threat. But even if we cannot change our systems of governance or the pervasive norms of the market overnight, it is clear that the faster we initiate changes the better. We are a species teetering on a cliff-edge. Either we start to manage our natural conformism, religiosity, and tribalism more effectively or we stampede over the precipice.

Our ancestors often had to make tough choices along these lines, but the consequences were usually less obvious. I sometimes wonder if the great monoliths at Göbekli, far from being the high water mark of an ancient hunter-gatherer world, were more like its final death throes. Perhaps the imagery of wild animals carved into mighty stone monuments was a desperate attempt to preserve beliefs in the mysterious forces of nature – a last-ditch effort to defend the old ways and values as they were starting to be questioned. At any rate, those who cleaved to foraging ways of life would eventually have died out, as successive generations were gradually seduced by new-fangled methods of cultivating crops and herding animals. The winners in this process were not the creators of the great monoliths but the farmers who gradually went on to build more widespread and enduring cultural systems during the Bronze Age and the great empires that eventually followed. Now, the world stands at a crossroads once again. But this time we have little excuse for failing to recognize which way lies survival and which way would mean stampeding into the abyss. If we continue to head in the direction of unsustainable levels of pollution and overconsumption of resources, we will face an environmental catastrophe.

Problems requiring large-scale coordinated action and forward thinking beyond our own lifetimes have been successfully solved in the past by developing more cooperative cultures and more future-minded religions and political systems. However, for humanity to find a path away from the cliff-face today, many of our most basic assumptions about ourselves and our societies need rethinking, and at a more rapid pace than ever before. As everyone knows, changing the direction of the human herd at the rate and on the scale required will depend on revolutionary changes to public policy, such as the introduction of carbon taxes, massive changes in the efficiency of agriculture and basic infrastructure, and the electrification of transportation systems. Anyone can see that the obstacles to transforming our societies in such fundamental and costly ways are formidable. But some of the most powerful obstacles are psychological, and these are also the most widely neglected.

Consider the human bias towards short-term thinking. Most of us already know that in the absence of massive changes to the way our economies function, the climate crisis will get much worse very rapidly – and have noticed too that action to prevent this is too piecemeal and slow. However, what we tend not to appreciate is quite how deep-rooted this tendency is in our evolved psychology. From the perspective of evolutionary psychology, we have inherited a natural predisposition not only to prioritize problems we face right now rather than those in an uncertain future, but also to be increasingly myopic and focused on the short term the harsher the environment becomes.

Moving the dial on these issues is likely to require more than just reason and evidence alone. In 2023, I sat at my computer screen in Oxford and recorded a series of speeches on video in which I argued that we must act as one to address the climate crisis. In each speech, the wording was the same – except for the bit in the middle where I appealed to a different logic to justify my conclusions. In the first speech, I used the standard arguments of science that are widely made in the media. But in the other speeches, I appealed to our evolved biases instead. Roughly a thousand members of the general public in the UK were invited to watch my little videos. After doing so, participants had an opportunity to donate money to an environmental charity. One of the key findings of this study was that appealing to notions of shared essence, but not to the arguments from science, was significantly more effective in motivating fusion with humanity and action on the climate crisis than hearing no speeches at all. In subsequent versions of this study in which the speeches were given by a prominent politician from the global south, the shared biology speech had significantly stronger effects on fusion with humanity than any of the other speeches.[3] Activists and leaders alike need to appreciate that reasoning based on scientific evidence and rational argument may not be enough to bring about the sweeping changes needed to combat climate change. Knowing what we must do and why is not sufficiently motivating to redirect the herd. We also need to change our cultural systems – and this means appealing to our natural biases in new ways and on a massive scale.

I will argue that the needed cultural revolution can only be achieved through a deep understanding of how ritual structures our everyday life, and how it might be transformed. In Part Two, I argued that routinization emerged with the spread of farming, enabling larger-scale group identities to spread and stabilize. In this chapter, after nearly 10,000 years of repetitive rituals, it is time to consider where they have taken us. It might seem odd to jump through time in this way. My point here is not that rituals have remained unchanged from the Neolithic until now. Clearly many forms of routinization are distinctively modern phenomena, utilizing amplifiers, radio broadcasts, and other relatively new technologies and operating on a scale that would have been unthinkable for early farmers. However, I would argue that the same basic effects of routinization have persisted through world history, from the first imagined communities of early farmers to the superpowers of the space age. The way we cooperate – or fail to cooperate – is still governed by the biases described in Part One and which were transformed and harnessed during the rise of large-scale societies in Part Two. These products of our evolutionary past – our modern inheritance – constitute something of a poisoned chalice. When combined with the technological capabilities of the modern world – and scaled up to the level of over seven billion people – our natural urges now pose an existential threat to our species. Arguably the most striking example of this is the way humanity's large-scale herding behaviours are driving us to the brink of environmental disaster. However, just as routinization may be our ruination, it could also be our path to salvation.

This chapter focuses on the way routinization in the modern world currently shapes the way we extract resources, create goods and services, buy and sell them in the form of commodities, generate profits, and invest. This system is known as capitalism. It has come to engender such widespread patterns of thinking and behaviour that it is often mistakenly assumed to be wired deep into human nature. However, when we look at this system through an anthropological lens, it turns out to be just another example of mass conformism – one that has been steadily leading to a climate crisis from which we may never recover. Recognizing and addressing the problem only becomes possible by viewing our current environmental problems in a much longer timeframe.

The unnaturalness of capitalism

The colonization of Papua New Guinea by European powers in the nineteenth and twentieth centuries profoundly transformed the lives of indigenous peoples. It imposed relative peace between formerly warring tribes. It introduced aid posts and medicines that healed the sick and saved lives. It provided access to steel tools that made light work of tasks that had previously required gruelling effort using only stone axes. And it supplied cheap sources of protein (e.g., tinned fish) and storable carbohydrates (e.g., rice) that became attractive alternatives to hunting animals and laboriously growing root vegetables. At the same time, however, it led to the imposition of law and order at gunpoint, forced people into harsh forms of indentured labour, and led to indigenous beliefs and ways of life being declared wicked and backward. One of the most distressing of all the unwelcome changes wrought by colonization was the destruction of the rainforest to make way for roads, logging companies, and commercial plantations.

In the process of trying to make sense of all this, the Baining people I lived with in Papua New Guinea had started to question many of their own ritual traditions. Some of the ways in which they described this to me were distressing to record. I was told, for example, that until Europeans arrived, people had regarded their customary forms of attire as perfectly adequate. But then they were taught by Christian missionaries that displaying naked female breasts was sinful. In that moment they learned to feel ashamed of their own culture. They accordingly abandoned many of their local customs, modified others, and in some cases adopted entirely new ones – including European clothes. Many saw this process as a path to salvation, a way of building a new relationship between the ancestral past and future redemption, but it was also traumatic and difficult. As one of my close friends in the community put it (my translation):

> We look back because we are lost in the world of the white man's knowledge. Recalling our ancestors, Adam and Eve, we see that their lives were good in Paradise. But they sinned and now we must all toil and suffer. So we ask ourselves in our confusion today: 'Where are we to go? Where is our home?' All the white man's knowledge has blinded us to what we once possessed. Today, the mission tells us that Jesus will come in sight of all of us – those of us who are alive and those who have died. So we ask ourselves: 'Who will bring back *our* ancestors? Many people seek knowledge but we . . . are the last to receive it. So, who will transform the old world into a new world?'[4]

During the time I lived in this community, a new solution to the dilemma was being worked out. The solution went beyond the original vision for the Kivung established more than two decades earlier. Yet many argued that it was consistent with the radical spirit of the early movement. The solution was to embark on a unique form of protest against modernity and globalization by casting off European clothing, returning instead to traditional ways of dressing, even while remaining committed to the Catholic-infused cosmology of the Kivung. Those who endorsed this return to traditional costume came to perform a host of rituals to help them reconnect with their ancestors. Many long-forgotten traditions from the pre-colonial era were resurrected but some were also consciously adapted to changed circumstances. All these processes were accompanied by deep reflection on the nature and significance of rituals, both ancient and modern, indigenous and European.

This was an effective way of rejecting the arrogance of the missions and the soullessness of capitalism. But it did not lead to the revival of all local traditions, such as tribal warfare and agonizing forms of male initiation. People chose a new way forward that allowed them to maintain peace, law, and order, and to develop a uniquely Kivung philosophy. It had two dimensions to it. One was rooted in sacred relationships with indigenous ancestors modelled by Koriam – a guru-like figure who was universally regarded as the original founder and spiritual father of the entire Kivung movement back in the 1960s. The other was grounded in hard-headed politics, led by a cadre of highly educated indigenous leaders determined to defend the rights of Kivung members and their rainforest from the destructive effects of extractive capitalism. The latter role was epitomized by Francis Koimanrea, who went on to become the Governor of East New Britain Province (the region in which the Kivung was established), championing the protection of the rainforest and leading the country's response to environmental catastrophes. The system of beliefs and practices these leaders helped to establish gave rise to a peaceful, law-abiding community on a scale that is rare in Papua New Guinea – a nation tragically held back by some of the highest rates of violent crime of any country in the world. Against this background, the achievements of the Kivung were all the more impressive.

So, why were followers of the Kivung able to bring about such extraordinarily cooperative achievements while other regions of the country were not? Part of the answer to that question is that they were prepared to question the basic assumptions of capitalism, drawing on their own history and traditions. But an even more remarkable part of the answer related to ritual. The rise of the Kivung in the 1960s helped establish a durable method of protecting the environment. The movement's leaders used routinization to spread and stabilize a set of beliefs and practices that both preserved and adapted ancestral ways of life, uniting the population in opposition to the environmentally destructive impacts of logging companies and mining operations.

The experience of the Kivung hints at a profound way of embedding environmentally sustainable beliefs and practices. While many of us note and lament the destructive impacts of globalizing capitalism, we seldom consider what we can do about it. It is not always obvious that centralized governments and capitalist markets are just one form of life rather than the only or the best kind for humanity at large. The importers of T-shirts, prisons, bulldozers, and sawmills seldom seriously consider what might be learned by listening to the ideas of indigenous peoples in the lands they forcibly occupy. The reason it is hard to question globalizing norms is because they are so widely routinized. At the core of everyday life in all the countries of the world today are a set of shared ideas about economic life that seem so inescapably ubiquitous, it is hard to imagine life without them. When rituals become this extensively routinized, the ways of thinking and behaving they engender can become so deeply embedded in the fabric of day-to-day existence that we no longer notice how strange they are.

Consider the following phrases: everything has a price; time is money; you need money to make money. Not only are such phrases hackneyed and familiar but the ideas behind them are so obvious as to seem beyond question. It was a very different story, though, for the people I went to live with in the rainforest of Papua New Guinea. In that part of the world, few objects of any use had a price. Houses were made from materials gathered from the surrounding forest. Foods were grown in local gardens or hunted and fished. Coffee and cocoa did indeed have a monetary value but no other use. People who wanted cash for crops might plant coffee and cocoa trees and sell their fruits to middlemen, but few observed what became of that produce after it was sold. It was clearly inedible in its raw form but nobody in the village drank coffee or ate chocolate, so the end products of the manufacturing process were mysterious and indeed entirely valueless from a local perspective.

Money obtained from cash cropping could be used to buy certain desirable goods – steel tools such as bush knives and axes or imported shelf-stable foods such as tinned fish and rice. But money was only spent in that way on special occasions – for example, when a knife finally snapped because years of sharpening had made the blade progressively narrower. The only frequent use people had for money was to place it in containers as gifts to their ancestors. In those special receptacles in the village's temples, the coins gradually accumulated. But it was not invested for profit. Such an idea was alien to the people I lived with. Instead, whenever the money built up it was taken to the Kivung headquarters and eventually used to support public works for the community as a whole – such as aid posts or schools. On one occasion it was simply given away as a massive charitable donation to assist faraway victims of a cyclone in Australia, despite the fact that the recipients of this gift were much richer, mostly well insured, and extensively assisted by domestic charities and government aid.

When we consider the material lives of humans over the millennia, the pursuit of economic growth as an end in itself only features at the very end of that history. We are not really natural capitalists; we are foragers. Humans everywhere will happily become engrossed in picking berries from bushes, a propensity that is nowadays callously exploited by the gambling industry using slot machines with images of fruits that need to be matched up to obtain a reward. Hunting instincts also run deep in humans, often in ways that are acted out not only by shooters and anglers who actually catch animals, but also birders and trainspotters who merely track their quarry without needing to go for the kill. But although we are natural producers and consumers, it's not obvious that we are natural hoarders. Our foraging instincts evolved in a world of perishable foodstuffs that could not be stored or carried, so an instinct to accumulate would have made little sense in that context.

As societies became larger, the human relationship with things became more complex. Farming forced us to settle down, and the production of surpluses created the possibility that wealth could be stored and stockpiled. But none of this led inevitably to the situation in which we now find ourselves. Because at the core of economics today is an altogether stranger idea: that humans are utility maximizers.

This idea, which originated in the nineteenth-century utilitarian philosophy of the British philosophers Jeremy Bentham and John Stuart Mill, proposes that good actions should maximize their contributions to wellbeing (their 'utility'). Since the economist Alfred Marshall placed this idea at the heart of economics in late Victorian and early Edwardian England, it has become fundamental to the way most economists think about human behaviour. In this framework, the more material goods we acquire, the more utility (or, at least, potential utility) we gain.

But there have also been debates about the generalizability and coherence of this idea, particularly among scholars who are seeking to understand economic behaviour in societies where monetarized markets are lacking. In my own discipline of social anthropology, for example, a discussion has rumbled on since the 1950s about whether the principle of utility maximization is present in pre-capitalist economies. On the one side are the 'formalists' who argue that it is, so long as utility is interpreted in a wide variety of ways – not just as selfish accumulation of assets but as being directed to social goals, such as fulfilling legal or moral obligations to others based on principles of exchange or charitable helping. On the other side are 'substantivists' who argue that the whole idea of utility maximization is ethnocentric – a way of thinking about decision-making in a capitalist market economy that doesn't make any sense in a traditional society oriented to subsistence production, gift exchange, or chiefly redistribution.

I have long wondered whether this presents a false dichotomy, however. My impression is that highly outcome-oriented, utility-maximizing tendencies are present in all societies – as are highly ritualistic and non-utilitarian ones. But in our society, the 'utility-maximizing' side of this equation has been dominant for too long, with potentially devastating environmental implications. And the solution is hiding in plain sight: lying in models that colonizers failed to learn from the colonized, and that the contemporary world as a whole has failed to learn from history.

The hidden rituals of economic life

Early in 2023, I found myself in a small boat surrounded by Pacific Island politicians on our way to the tiny island of Iririki, just off the coast near the capital of Vanuatu. We were there in the wake of two major cyclones which had directly hit the country causing massive damage to infrastructure and leaving many houses without roofs, food supplies, or running water. The event we were attending had been organized by Vanuatu's minister for climate change, Ralph Regenvanu, who was leading a pathbreaking effort to demonstrate how quickly Pacific Island nations could phase out fossil fuels. It resulted in a call for a just transition to a fossil-fuel-free Pacific, arguing that the climate crisis was 'driven by the greed of an exploitative industry and its enablers'.[5]

The politicians and activists I met on Iririki gave me reason to hope that the tide might finally be turning on the one-way flow of globalizing influence from the rich countries to the poor. Less than two weeks later, Minister Ralph Regenvanu was at the UN headquarters in New York asking the general assembly to vote on whether the International Court of Justice should issue a legal opinion on climate responsibility that would establish the legal obligations of the world's nations to protect the people and ecosystems most severely affected by climate change. This was an historic moment. All previous attempts to get the ICJ to consider issues of climate change had fallen on deaf ears due to lack of diplomatic support but this time, at long last, the effort succeeded. The resolution was co-sponsored by 132 countries and adopted by consensus. The islands of Vanuatu are home to many indigenous language groups, all of which – by extraordinary coincidence – can trace their ancestry to the Baining people I lived with many years before in Papua New Guinea.[6] Like the Baining, who helped to devise methods of defending their rainforest home from the destructive incursions of capitalism,[7] their cousins in Vanuatu were now mobilizing on an international stage to lead the global struggle against climate change.

At the heart of the clash of cultures between commercial interests and indigenous groups are contrasting theories of economic life. Colonists and corporations are like the formalists in anthropology who see the world through the lens of market economics. Indigenous groups are like the substantivists who view economies as embedded in wider social systems. But the whole debate can be radically reframed if we bring to it the crucial distinction between ritual and instrumental stances that I first described in Chapter 1. To recapitulate briefly, we adopt an instrumental stance when our focus is on how behaviour contributes rationally to an end goal; we adopt a ritual stance when we assume that there is no underlying causal structure to the behaviour, and we are focused instead on reproducing established conventions and fulfilling social obligations. The former is motivated by the desire to achieve material results, the latter to affiliate with others.

Once we recognize the presence of both stances in human social life, the positions of formalists and substantivists become easier to reconcile. Formalists are right that instrumental reasoning drives utility maximization in all human societies, not just capitalist ones. But substantivists are right in claiming that rational utility maximization is not the goal of all economic behaviour. An appreciation of the ritual stance allows one to realize that not all our behaviour is acquisitive and materially oriented – and that much of it is instead driven by an urge to affiliate with the group.

Yet if both ritual and instrumental reasoning appear to play a role in economic decision-making in all societies, the balance between the two clearly varies considerably. For example, in many traditional societies, lacking all-purpose currencies and commodity markets, the balance between ritualistic and instrumental motivations for economic activity may be more in favour of the former. That is, the motivation to produce goods and to participate in systems of exchange and redistribution may be strongly influenced by conformism, tradition, and respect for ancestors in ways that are fundamentally about maintaining long-term social relationships. By contrast, in a modern city, economic transactions involving the procurement of food from supermarkets or the generation of profits from customers and clients may be ephemeral and shallow – based on instrumental reasoning rather than the desire to deepen and extend social relationships or maintain institutions.

While it is tempting to place economic systems on a single continuum – more ritualistic (e.g., traditionalist) at one end or more instrumental (e.g., capitalist) at the other – the reality is more complex. It could be argued that most, if not all societies, are simultaneously constructed around at least two kinds of economic life oriented respectively to the procurement of things of material value and things of social value. A simple illustration of this is the *kula* ring – a system of exchange described in the Trobriand Islands of New Guinea by Bronislaw Malinowski, one of the founders of social anthropology. *Kula* valuables – in the form of beautiful armshells and necklaces – were traditionally traded across Pacific islands in opposing directions, creating a vast ring of exchanges in necklaces travelling clockwise and armshells anticlockwise. As boats travelled between islands, they brought with them not only *kula* valuables that could be exchanged but also useful goods that could be bartered. Barter was focused on the procurement of useful goods at the best rate possible – as one might expect in any form of market exchange. The motivations of producers and traders was mostly quite instrumental in orientation – focused on getting the best deals relative to effort invested. By contrast, the more publicly visible and ceremonially elaborated exchange of *kula* valuables was the social highlight of every voyage. The motive for *kula* exchange was to acquire social status by strategically giving one's most valuable objects to the owners of ones of greater value, compelling the latter to give up their more valuable objects in return, or to lose face. Individuals became famous throughout the region as the owners of the most prestigious *kula* valuables, even though it was difficult to hold onto them. This system was dominated by the ritual stance, aimed at forging relationships between givers and receivers and ultimately cementing bonds of alliance between islands based on normative rules laid down by universally accepted tradition.

The balance between ritual and instrumental stances in the economic lives of Trobriand islanders may be difficult to quantify precisely but both perspectives were pervasive features of exchange, albeit in varying degrees. In barter, the aim was highly utilitarian: to obtain materially useful goods for consumption. Whereas in *kula* exchange the aim was much more ritualistic: to obtain prestige goods and hold onto them for as long as possible in order to maintain connection to a glorious past. But there was a difference of emphasis. In the world of barter, the focus was on more purely instrumental thinking aimed at meeting material needs rather than building strong social relationships. But in the world of *kula* exchange, elaborate protocol was more pronounced, and the emphasis was on affiliation, alliance, and political integration rather than merely on the consumption of goods.

Herein lies the hidden key to understanding economic behaviour in capitalist systems – which of course most people (including Trobriand islanders) are immersed in today. Viewed through the lens of a business model, behaviour in the marketplace is essentially materialistic, driven by the instrumental stance. But this is only a one-sided view of our behaviour. If you sidle up and view the world of economic life instead through my anthropological lens, it looks quite different. From that perspective, the logic of the ritual stance is strangely ubiquitous in even the most capitalistic systems. For example, even a casual observer of the English class system will appreciate that at its heart resides a logic very similar to that of the *kula* ring.[8]

For the English aristocracy, the equivalents of *kula* armshells and necklaces are land and artworks associated with prestigious lineages, ideally of royal association or bloodline. All such objects can and do change hands via mechanisms of inheritance, appropriation, or conquest. Just as Trobriand chiefs sought to become associated with the most prestigious *kula* valuables and their legendary former keepers, so the *nouveau riche* everywhere seek to cloak themselves in an aura of noble ancestry. For example, in Oxfordshire, where I live, there is a long tradition of newly monied financiers from the City of London buying up the most historically significant houses once owned by aristocrats. Even the most calculatingly profit-maximizing individuals, it seems, are not immune to the status brought by possessions with illustrious pasts. In much the same way, multinational corporations compete to adorn their boardrooms and lobbies with the works of famous artists. And just like the Trobriand chief, the goal of the capitalist in the prestige economy of art-collecting is to acquire objects that confer status and legitimacy by association with tradition.

Even if we accept that the economists' idealization of the rational actor in a capitalist society is mostly driven by wealth creation, then, it is clearly not the whole story. It may be that accumulation is the utility that business leaders care about most but, if so, it is not their only motivation. Nor is it the most natural or intuitive goal for human beings to have. The difference between the most 'instrumental' capitalist societies and the most 'ritualistic' gift economies is therefore only one of degree. And with that historical and ethnographic background in mind, we can begin to see the capitalist way of viewing and interacting with material objects in a somewhat different light. To my mind, the most fundamental implication of this insight is that the dominance of materialistic culture at the core of capitalism is neither natural nor God-given – and, indeed, altogether less 'instrumental' ways of being are always just beneath the surface.

And that means our lifestyles and goals can be changed. In fact, we could replace them with a quite different set of habits which, with practice and effort, could soon became established as a 'new normal' – a set of ways of thinking about economic life that become as deeply embedded in daily life as any of the routinized rituals that now dominate our systems of production, consumption, marketing, and exchange. Indeed, this is something I have actually witnessed first-hand, as I now explain.

Conformism and capitalism

The way the Kivung addressed the problem of how to defend their rainforest from the ravages of capitalism was as brilliant as it was simple. Its leaders established a highly routinized system of monetary donations. On one level, the goal of this was purely instrumental. Kivung members set out to build up a centralized pot of money via a system of taxation, which could be used to fund the movement's environmental goals, from supporting the election of its leaders to positions in provincial and national governments to attracting forms of foreign investment in line with Baining aspirations to protect their ancestral lands. On another level, however, the goal was understood in terms of an elaborate moral framework, in which donations were interpreted through the lens of the ritual stance. Each donation was addressed to the group's ancestors, in much the same way as offerings of food were laid out in temples. And the whole idea of attracting foreign investment was interpreted at a deeper level by Kivung members to mean that their donations would serve to persuade the ancestors to return. The notion of 'foreign investors' was just code for ancestors. When the 'investor' ancestors returned they would bring with them the political and economic clout necessary to fend off the destructive effects of capitalist exploitation. This new heaven on earth would be one in which the Kivung faithful would live a life of luxury and ease. There would be no suffering, sickness, or death. And outsiders bent on stealing from them their land and resources would be punished.

The standard reaction of many outsiders to the Kivung was to write it off as a delusional belief system pinning all hopes for the future on a supernatural intervention that would never happen. However, this overlooked a much deeper and more important truth about the Kivung. It provided a potent motivation for its rank-and-file members to contribute to a new political and economic system that was devoted to protecting the ancient rainforest. The rest of the world can learn from their example – not by spreading some millenarian vision but by looking beyond Kivung beliefs about returning ancestors to the more fundamental goals of the movement and its methods of implementing them. What the Kivung achieved was a way of combining routinized ritual and instrumental thinking into a potent collective action plan to address pressing environmental issues. This is precisely what the capitalist marketplace alone struggles to accomplish.

After all, the capitalist world has no shortage of ideas about what a better economic system would look like. Take the notion of conscious capitalism, which suggests that wealth creation itself should be a process in which prosocial goals are present from the get-go.[9] Or consider the similarly novel set of proposals for rewriting economics that comes from my colleague in Oxford, Kate Raworth. Like the anthropological substantivists of the 1950s, Raworth argues that economies need to be understood as embedded in larger social systems. But this is not only true of traditional gift economies; it is true also of commodity markets. Raworth posits that mainstream economics – which emphasizes the maximization of Gross Domestic Product (GDP) via market exchange and the minimization of government interference – is both blinkered and unsustainable, because it ignores the dependence of our economies on finite nonrenewable resources (not to mention hidden costs such as infrastructure and child-rearing). The solution, Raworth argues, is to imagine the world economy as a doughnut – the kind with a hole in the middle.[10] The inner ring of the doughnut is the minimum for a just society: one that meets people's provisioning needs for food and water, housing, energy, employment, public health, education, peace, and so on. The outer ring represents a set of ecological constraints which, if transgressed, would imperil the very basis for life: things like a stable climate, a protective ozone layer, sufficient freshwater, clean air, healthy oceans, and so on. The ideal place for any economic system to be is somewhere in the chewy dough. It should not trample recklessly onto the surrounding space by fouling up the oceans, punching holes in the ozone layer, or triggering climate change. But nor should we fall through the hole in the middle by coming up short on the provisioning of basic needs.

So, it seems to me that the issue we face is not a lack of visionary thinking about what a better economy might look like in theory – but rather the thornier question of how to actually embed these lofty ideals in practice. One of the challenges faced by freethinkers like Raworth is that of persuading business leaders to break with the long-established norms of the marketplace. At its heart, capitalism is highly ritualistic. Its cardinal values are underwritten by routinized rituals, subtly at play in boardrooms and business meetings but which are rooted in fears of ridicule and exclusion that emerge early in development, even before children go to school. Recall that even at a very early age, humans yearn for acceptance by the group and fear exclusion. Businesspeople are no different. What will others think of them if they start to pursue goals other than the conventional measures of success: market share, sales revenue, return on investment . . . ?

And yet if routinization is at the root of our problems, routinization must surely also be part of the solution. Just as followers of the Kivung were able to transform their collective behaviour by adopting new rituals, so too can we. Consider how many times efforts to bring about changes in collective behaviour have succeeded when many thought it would be impossible, even in living memory. When I was an undergraduate in the 1980s, it was normal to smoke in the classrooms at my university as well as on buses, trains, and aeroplanes. Racist and openly fascistic chants rang out in the stands at football matches. The prospect of gay marriage seemed unthinkable. In some such cases – here in Britain for example – there has been a striking change of habits, such that once the new behaviours had become sufficiently widespread and routinized, it rapidly began to feel as if things had always been that way. If someone were to light up a cigarette in a classroom or on a bus in my home town of Oxford today, those around them would be shocked; it takes a real effort to remember a time when that would not have been the case.

Much of the work required to think differently boils down to adopting new rituals and norms. Such new rituals could be spread by trendsetters within individual businesses and organizations. For example, imagine if restaurant menus routinely included a carbon footprint estimate alongside calorific values, or if schools were required to teach all children about the effects of consumer choices on climate change? Or what about if religiously based food taboos – such as prohibitions on eating pork – were to be reframed around environmental concerns, such as a prohibition on eating beef, which of course is one of the more damaging causes of carbon emissions on the planet? What if, given the urgency of the climate crisis and its global implications, all major festivals and ritual celebrations around the world were redesigned to incorporate elements of pro-environmental action?

It is important to remember that ritual routinization is by no means a uniquely religious phenomenon. Soviet communism similarly introduced a plethora of repetitive revolutionary festivals, memorial celebrations, and special holidays, and ritualized many aspects of day-to-day domestic life to promote loyalty to the state.[11] The Chinese Communist Party likewise established many new rituals to replace more ancient folk traditions – including the chanting of standardized slogans and synchronous marching in military parades. I recall still hearing these public announcements as I travelled around China in 2008. Eerily evocative of the call from the minarets of Islam, public broadcasts through loudspeakers ensured that every morning and evening my day began by listening to revolutionary songs. In this way, the Communist Party effectively systematized folk religious practices as part of the new ideological framework embedded into the habits of daily life.[12] Nor has modern western society been immune to this seductive logic. Although the effects may be more subtle, consider how music piped into clothes retailers in shopping malls and on the high street attempts to influence our consumer behaviour when we are choosing to buy new clothes and bodily adornments. But what if secular institutions today were to use these same techniques to routinize collective rituals to create more sustainable lifestyles? Even if we cannot imagine the governments of liberal democracies piping environmentalist slogans Soviet-style into our public spaces, maybe we can imagine such shifts in messaging in our entertainment industries.

Celebrities and entertainers exercise extraordinary influence over conformism: the readiness to follow what other people do not because of its instrumental benefits but because of the desire to be like them. From the perspective of the climate crisis, this might mean going beyond the instrumental arguments that come from science and paying just as much attention to the people we want to imitate.

If you are religious, it might be your spiritual leaders; if you are a football supporter, it might be the star players in your team; if you are a fan of popular music, it might be a singer. Currently only a smattering of our celebrity role models are worried enough about our global environmental crisis to mobilize their followers and imitators to care about it too. However, despite the fact that such celebrities occupy very diverse moral and cultural landscapes, they are eminently capable of helping us to coordinate around common goals. Why is there no well-established institutional forum for exploiting such potentially influential positions in our societies – no equivalent of canonization, World Cup trophies, or Grammy awards for celebrities leading action on the climate crisis? Why is there not even a sustained effort by thought leaders and media providers to push, cajole, or otherwise induce the people we most seek to imitate into taking that responsibility seriously?

Such ideas perhaps sound somewhat optimistic. In theory, however, they could be implemented quite swiftly, with much the same kind of inspiring leadership that once produced mass civil rights movements or Live Aid concerts. Currently there is a vacuum for such activists waiting to be filled. There is now broad consensus that the climate crisis requires urgent action and that changes in behaviour on a massive scale, particularly in wealthier countries, need to be part of the solution. Much of the battle is therefore already won. However, changes in our collective habits of thinking and behaviour must also be catalysed by courageous leadership at all levels of society – but particularly among those who have the privilege to be cultural role models – to coordinate action and herd us in the right direction. Even if our vote-seeking politicians are mostly followers of public opinion rather than leaders of it, our cultural icons can afford to take more risks. Moreover, the risks may be less than they appear. Most of us alive today are participants in global capitalism and in this respect, we are all herding together. But the herd is spooked. We are ready for a change of direction.

Ritual and the rebirth of future-mindedness

If there had been paparazzi in traditional Melanesian chiefdoms, you would have found their flashing cameras at the publicly ostentatious exchanges of armshells and necklaces: the crowning moments of the Trobriand islanders' *kula* voyages. But in the background, much less obvious and more mundane business transactions were taking place. These focused on the distribution of useful goods around the islands rather than on the great art objects changing hands, the speeches being made, and the impact of the ceremonies on inter-island alliances and obligations. This prosaic economic activity was primarily instrumental in its material goals and motivations. And such transactions were therefore also based on short-term thinking. Goods changed hands according to present needs, and not with an eye to longer-term goals. With the exchange of *kula* valuables, it was the other way around – all gifts and counter-gifts were calculated to maximize opportunities in the future to create obligations. In time, these would force exchange partners to hand over the prestigious possessions with the most illustrious histories, forging robust and enduring mutual obligations and political alliances among far-flung islands. Such future-mindedness was essential to the peaceful coexistence of chiefdoms.

Barter and capitalism are similar in part because both tend to be motivated by relatively short-term, selfish strategies rather than by future-minded and prosocial concerns. True, the pursuit of profit in a capitalist economy involves planning and investment for the future – but the aim is typically to drive forms of growth that are unsustainable and therefore poorly adapted to future flourishing. The adoption of an egoistic instrumental mindset linked to personal gain – rather than a ritual mindset linked to group belonging and prosocial outcomes – is arguably making us more short-sighted.

Understanding this could have important consequences for the way we manage the global environment. Given the scale on which irreversible damage to the planet can be expected to occur in the decades and centuries to come thanks to decisions made today, the short-sightedness and hesitancy with which governments have been responding is alarming. The crunch question, then, is how we might hope to reduce short-termism and strengthen resolve to adopt new ways of life *en masse* and on a global scale. Here, too, routinized ritual might offer the solution. In Chapter 4, I argued that in past civilizations, frequent repetition of the group's customs, beliefs, and practices helped our ancestors to become more future-minded. We encountered, for example, the history houses of Çatalhöyük in which people came to see themselves as part of a much deeper tradition that would continue long after their own short lives. And we considered how increasing the pace of ritual life would have encouraged a more future-minded outlook.

This is not mere conjecture. In fact, we now have firm experimental evidence which shows that viewing collective life through the lens of the ritual stance on a regular basis improves our self-control and makes us better able to defer gratification. Our main aim was to find out whether the process of being socialized into a system requiring participation in routinized rituals would make people better able to prioritize greater longer-term rewards over lesser short-term rewards. We agreed it would be important to compare people in both a modern-facing western country and a more ritualistic traditional society, so that we could establish whether the effects of our experimental manipulation would work regardless of the background environment. This took us to Vanuatu, where many people continue to follow traditional lifestyles.

Our starting point was that undertaking a routinized ritual would involve more concentration and self-regulation than simply acquiring a new instrumental skill. Learning a skill involves focusing on the end goal and seeing how the steps involved help to get us there. As such, you only need to copy the things that help to produce the desired outcome – and can ignore everything else. Rituals are different. Recall that when we adopt a ritual stance on behaviour, we feel compelled to do everything the person we are copying does, especially if the actions appear to be deliberate. This requires much more concentration. It means one must attend closely to every little detail – you can't discard the seemingly 'pointless' elements. Even more onerously, you must retain that level of attentiveness until the action sequence has been thoroughly learned. We hypothesized that the self-discipline required to learn a ritual – and repeat it over time – would make people less impulsive than when they were learning a new practical skill. In this way, routinization might lead in turn to greater future-mindedness.

If correct, the benefits of participating in routinized rituals might be most substantial and lasting if we started young, by trying to influence the development of minds still forming. So, my student – Veronika Rybanska – recruited roughly a hundred children around seven years of age in each country – Vanuatu in the Pacific and Slovakia in Europe – to take part in our study over a period of three months while attending school.[13]

Before the study began, measures of self-control and ability to delay gratification were measured for each child. The first measure was based on a version of a 'head-to-toes' task that psychologists commonly use to measure children's ability to deliberately control their movements. The participating child was given a series of rapid instructions: 'Touch your head!', 'Touch your toes!', 'Touch your shoulders!', or 'Touch your knees!' When it was clear that the child had mastered the ability to carry out these instructions correctly, they were told to act in ways that contradicted the commands in a systematic way – for example, to touch their shoulders when instructed to touch their knees or to touch their toes when instructed to touch their heads. The child's ability to carry out these intentionally misleading commands according to the agreed rules provided a measure of executive function. Meanwhile, to test their ability to delay gratification, children were given a sweet and told that if (and only if) they could wait patiently and not touch it until the experimenter returned, they would receive two pieces. The experimenter then left the child alone with the treat for up to fifteen minutes. If the child touched, smelled, or tasted the sweet, the test ended and the number of minutes they had managed to resist was recorded.

The children duly tested, they were then divided up and assigned to one of three groups which they would stay in for the next three months: a ritual condition, an instrumental condition, and a control condition. Children assigned to the ritual and instrumental conditions were taken out of class twice a week in groups of eight to ten to participate in so-called 'circle games', originally designed to improve children's ability to self-regulate. During each of the sessions, children participated in six games, lasting in total for between thirty and forty-five minutes. In the ritual condition, children were given reasons for the games that emphasized the importance of following the rules and the conventional nature of the activity: for example, reminding them that 'it has always been done this way'. By contrast, in the instrumental condition, children were given reasons for the games that emphasized their practical benefits and pedagogic end goals: reminding them that 'if we do it this way we will learn about different animals'. The children assigned to the control group were given no opportunity to participate in the circle games at all. After three months of repetition, the children's executive function and ability to delay gratification were measured once again.

Although experiments in field settings are difficult to implement and often go wrong, this one produced a clean set of closely matching results in both countries. As predicted, children in the control condition showed the least improvement in either executive function or ability to delay gratification. Children in the instrumental condition fared much better. But the children who benefited most from the intervention were those in the ritual condition: the ones who had been told not to focus on what they might learn, but just that 'it has always been done this way'.

We had our first clear experimental evidence that, even among highly secular European children, ritual participation promotes self-regulation and future-mindedness. And this was a discovery with startling implications. If our routinized rituals fall into abeyance, our ability to defer gratification will be impaired and our orientation to the future will be more influenced by short-term thinking. This would suggest that societies currently shedding their rituals and traditions most rapidly are also eroding their capacities for long-term thinking. Consider for a moment what this means for processes of secularization. According to Pew survey data from 2014, each generation of Americans born since the 1920s participated less in routinized religious rituals than the one preceding it. Those born before 1945 were most likely to pray daily and attend religious services at least weekly in 2014. Baby Boomers were less frequent participants in both types of ritual, members of Generation X less frequent still, and Millennials the least frequent participants of all.[14] These patterns are by no means unique to the USA. Research shows that young adults in many countries report lower rates of affiliation with organized religions compared with their elders.[15] While the decline of ritual may not be the sole cause of our pandemic of short-termism, it seems unlikely to be helping.

On the other hand, might a return to routinized rituals help us become more future-minded once again? Some suggestive evidence comes from research into the effects of regular attendance in church services on homeless people, and especially patterns of substance abuse.[16] For example, in one study of 380 homeless individuals in Canada, those who participated in church services more frequently were found to have significantly lower rates of alcohol, opioid, and cocaine use than less frequent participants. One possible explanation is that participation in routinized rituals encourages drug users to be more concerned about the future and therefore the harmful long-term effects of substance abuse, instead of the more immediate pleasure of their next hit. This is consistent also with the finding that frequency of church attendance in America is associated with charitable giving and volunteering for both religious and secular causes – suggesting elevated concerns about the future, whether in this world or the next.[17]

It would not be unduly far-fetched to see our reliance on fossil fuels and excessive meat consumption as forms of addiction. But as with those drug users, a solution to our dependence may lie in the reintegration of routinized rituals into our consumer behaviour. This is not an uncontroversial notion. Among environmental activists, the idea that consumers can change the world is polarizing. Some would argue that placing the burden of responsibility on individuals rather than corporations and governments is a cynical ploy to distract attention from those who should bear the costs of greening our consumption habits.[18] For example, reframing the damage to the environment caused by fossil fuels as a problem that can only be addressed by consumers using their motor vehicles more responsibly, or holidaymakers choosing shorter-haul destinations, enables the oil companies and airlines to continue marketing and pricing strategies that incentivize overconsumption in the interests of their destructive industries. Politicians also may be eager to pursue policies that shift responsibility onto citizens to reduce their individual carbon footprints, in order to deflect attention from the need for the less popular or even politically suicidal forms of public policy that are actually required. But I disagree, at least in part. While structural change is indeed required, it will not happen without massive changes in both norms and behaviour on the part of the majority. Greenwashing only works as a strategy as long as it does not obstruct the goal of polluting industries to make us consume more of their products. Such industries have everything to fear from changes in our individual mindsets that lead to widespread behaviour modification on the ground. The reason corporations like BP can afford to pour money into ad campaigns supporting greener consumption behaviours is that instrumental thinking has only a weak effect on behaviour (think health warnings on cigarette packets). By contrast, powerful vested interests in industries contributing the most to climate change should rightly be afraid of large-scale people power rooted in conformism, religiosity, and tribalism. It is these resources of human nature that could finally turn the tables on the polluters.

A potentially game-changing place to start might be to change our eating habits. Commensality – the simple act of eating together – has been a core feature of domestic ritual for many millennia. The first signs of its decline in modern times became apparent with the advent of so-called 'TV dinners' and has reached new heights with the rise of online entertainment fed to us through smartphones, tablets, and computers. Could something as simple as the resuscitation of collective mealtime rituals help us become more future-minded, and in turn become more focused on the future of our planet? Eating together provides seemingly limitless opportunities to build new forms of routinized ritual: ones that serve the interests of society as a whole and planetary health in particular. School dinners would be one of the most obvious environments not only to offer greener options, but also to routinize collective eating habits, perhaps even integrating consumption and learning environments by making canteens into effective classrooms and ceremonial sites that support green agendas and innovations. But this need not stop in childhood. Our lives abound with potential sites for commensality, from restaurants to workplace cafeterias.

New technologies might also be used to reintroduce routinized rituals into our lives. For example, smartphone apps could be used to help gently direct consumers towards more future-minded purchasing decisions in everyday life. The science of environmental nudging has identified scores of interventions that successfully change behaviour in ways that reduce our harmful effects on the environment. Indeed, one recent metanalysis concluded that repetitive digital 'nudges' for pro-environmental action are 'incredibly successful', causing participants to modify their behaviour by taking shorter showers and optimizing heating and electrical consumption.[19] Such interventions could take any number of forms, however. Consider the example of MyEarth, an app that tracks individuals' carbon footprints and gives them daily reminders of the effects their actions are having. This approach is endlessly scaleable: imagine an app that instantly calculates the relative benefits of buying a cheap product with a shorter estimated lifespan and a higher-quality product with a lifetime guarantee. If this shifted consumer behaviour towards longer-lasting products, it would gradually herd our throwaway culture in a more ecologically sustainable direction. Crucially, however, the effects of nudging can only be expected to endure if the behaviours they trigger become embedded through routinization, as I argued in Chapter 4.

As technology becomes more sophisticated, so too do the possibilities of routinizing its more beneficial consequences. Imagine if your regular food bills could be used to provide weekly or monthly feedback on the environmental impact of your purchasing patterns over time, or your performance in relation to consumption targets you had set yourself in advance? Such data could be used to generate automatic projections of how your consumption patterns would affect the environment over years or decades. Simulations could be run to show how such effects would scale up if they were typical of your local community or if they were adopted in the form of national or global consumption habits. The possibilities are endless.

As we have seen, inducing routinized ritual takes effort, particularly in a world that is rapidly secularizing. Educators therefore also have a potentially important role to play in helping to routinize sustainable consumer habits as part of the socialization process. Prioritizing quality and durability over immediate benefits based on price and convenience is a skill that cannot be acquired easily without support. And this is where the encouragement of policymakers is required. If one of the key functions of formal education is to cultivate habits that counter the harmful effects of advertising, then sustainable routines of consumer behaviour might reasonably be seen as core to a rounded curriculum.

Consumption habits are part tradition and part skill. They are collective patterns of behaviour acquired through a combination of imitation and past experience, influenced by role models and information providers. Do we want those models and providers to be industries and advertisers with vested interests orthogonal to those of consumers? Or should at least some of the responsibility for imparting our consumption strategies lie with caregivers, communities, schools, and governments? If the latter, consumption habits and skills might start to look very different.

In the process of colonizing Papua New Guinea, Dutch, German, British, and Australian government officials and missionaries set out to enlighten people they considered to be primitive. It seldom occurred to them that they could themselves benefit from the wisdom of indigenous groups. However, the lessons I learned from followers of the Kivung are of incalculable value not only to rich and affluent countries that have already profited from resources plundered from the poorer ones, but to humanity at large.

Much of this chapter has been given over to these lessons. For example, it was from the Kivung that I first learned the importance of routinization as a means of changing behaviour at scale and fostering a more future-minded approach to collective action. But perhaps the most important insight I took from the Kivung was that we can collectively challenge the hegemonic norms and power structures that tower over us. When the Kivung first became established, people already knew that they stood at a crossroads. People could, if they wished, follow the ways of colonial settlers by dressing in European clothes, attending church, and establishing commercial businesses such as logging companies or plantations of copra, coffee, or cocoa. The attractions of this pathway were obvious. They included access to desirable goods, schools for their children, and the promise of medical care for the sick and elderly. Even more compellingly for some there was the prospect of affiliating with the Europeans and working together with them in business, government, and public service. But there were also some obvious downsides to such an approach. The kinds of businesses that could realistically be established were mostly destructive to the rainforest on which people's lives had always depended. The damage wrought to the environment by logging companies and commercial plantations was already painfully apparent and establishing more of the same would only worsen the problem. Furthermore, admittance to the club of colonists and missionaries patently failed to deliver the forms of equality that indigenous people deserved.

So followers of the Kivung met to come up with a better plan. And to do so, they adopted a profoundly democratic process. Entire communities gathered to discuss the pros and cons of various policies and they did so, not just once or twice, but on a very regular basis. In fact, the word 'kivung' in Tok Pisin (the most widely spoken trade language in Papua New Guinea) means quite simply 'a meeting' or 'to meet'. The deliberations of Kivung followers were undertaken in over a hundred meetings a year in every village of the movement, scaling up to many thousands of meetings across the movement as a whole. These gatherings were governed by a framework inspired by the Ten Commandments of the Old Testament, and during twice-weekly meetings in every village, people discussed each of the ten laws in rotation – starting again at the beginning after each five-week cycle – allowing them to explore the implications of each law for a huge variety of challenges faced by their communities and to arrive at highly creative and consensual solutions. In their way, these meetings were rather like the 'citizens' assemblies' found in a handful of liberal democracies:[20] bodies in which representative samples of a country's general population come together to debate problems, informed by expert advice, and come up with solutions – ones that may often be far better than those arrived at by parliamentarians.

Through this deliberative process, the Kivung movement came up with a winning strategy. Adherents concluded that the peoples of the rainforest should join forces with their colonial overlords but only on their own terms. Since joining the capitalist system would have meant degrading their environment, they decided to restrict involvement in cash cropping and to implement a collective prohibition on land sales to logging companies and plantation owners. Likewise, they would greet the visits of colonial patrol officers in a friendly and cooperative spirit, but they would privately handle all problems of crime and conflict internally, leaving little for the state police to worry themselves about. By maintaining this dignified aloofness from the system of colonial administration on the ground, indigenous communities would protect themselves from its worst forms of bullying and exploitation. And finally, they would endeavour to infiltrate the upper echelons of government by establishing their leaders in prominent positions in both national and provincial assemblies by harnessing the electoral power of Kivung members as a way of voting *en bloc*. Their approach was highly successful, enabling communities joining the movement to maintain law and order, to protect their land more effectively against the incursion of outside commercial interests, and to gain representation of their collective interests in government.

This focus on collective decision-making is not to say that the Kivung did away with leaders altogether. On the contrary, every village had its leaders who led the discussions at regular community meetings and the structure of the Kivung as a whole was both hierarchical and centralized. But the organization was set up and practised in a way that encouraged collective planning through open discussion and debate.

Our future, too, will depend on the interplay of our systems of decision-making and the kinds of leaders we choose. We will need scientific leaders to provide our electorates and citizens' assemblies with reliable information on the mechanics of cause and effect – problems of an instrumental kind. But we will also need leaders we can trust to help us solve problems and implement solutions that relate to human relationships and the moral fabric of society. The Kivung brought forth leaders of both kinds: a guardian of tradition in the form of Koriam who communed with the ancestors and helped people to interpret the ten laws but also a younger cadre of highly educated leaders like Koimanrea who were capable of navigating the potentially exploitative world of politics, government, and commerce. What these leaders had in common, however, was the courage to speak up for the interests of their communities, even when it meant telling people inconvenient truths and winning them over. The trust and respect this required was also derived from consensus-building and routinization. And that in turn required courage on the part of followers: to back ideas that were sometimes challenging or surprising, to run the risk of exclusion and ridicule from those who didn't join the Kivung, and to herd in a novel direction simply because they believed in it. This is the kind of courage we must all now have if we are to overcome the threat posed by the climate emergency.

8.
Religiosity on Sale

I found myself surrounded by thousands of people, many of them with hands raised, gently swaying from side to side to the music. We were all transfixed by the floodlit stage on which singers with microphones were moving in synchrony. To the right of the stage was a large band of musicians – drummers, rhythm and bass guitarists, a brass section – and to the left of the stage a choir, perhaps twenty strong. Occasionally I caught glimpses of technicians scurrying around behind the scenes, moving equipment and props. But most eyes were on the giant screen above, where the faces of leaders with microphones were illuminated in perfect detail and being beamed around the world. This was not a rock concert. It was a Christian church service on a regular Sunday in Singapore.

Click on the website of New Creation Church and you immediately see images of attractive and seemingly joyful and affluent young people smiling at you and at each other. The images blend with slogans saying: 'Welcome to our family', 'We're here for you'. And there are ads for New Creation television: 'Broadcast the gospel to the world that all may see the loveliness of Jesus and the perfection of His work.' As I glanced around the vast auditorium, I noticed that many people were indeed young and attractive like those on the website, dressed in casual western clothing displaying fashionable brands. Everyone had sipped the wine and tasted the wafer representing the blood and flesh of Christ. But there were far too many people here – in the tens of thousands – to form an orderly queue before the pastor. Instead, they had been provided with the holy sacrament in their seats via small capsules distributed around the audience by countless volunteers. They resembled the pods used in coffee machines but with a peelable top containing the wafer and a plastic section beneath containing wine. Thousands of these mass-produced holy morsels had been consecrated at a single stroke.

The auditorium in which the congregation had gathered spanned the upper floor of a huge shopping mall, purpose-built to enable visitors to purchase and worship in a single trip. From the glitzy entertainment to the turnstiles and ushers showing us to our seats, everything about this environment bore the hallmarks, and the trademarks, of commercialization. It struck me how many people now treat religion in much the same way as a shopping mall. In the case of Singapore, roughly three-quarters of the country's citizens identified as ethnically Chinese, most of whom were Buddhists, and the rest largely a mixture of Christians and Taoists. The remaining quarter of the population were nearly all either ethnically Malay or Indian and so there were also sizeable numbers of Hindus and Muslims in the country. Add to this the fact that each of these religions came in multiple varieties and it was clear that the number and diversity of products in the religious shopping mall were quite staggering. Moreover, Singaporeans like to shop around. Many of those singing and swaying next to me in this vast congregation also observed traditional Chinese beliefs and practices with their families, including the veneration of ancestors. Some also regularly attended Buddhist temples.[1] Several of the people I talked to came to Christian church services partly in order to improve their English or to help their children learn.

As I joined the crowds streaming out of the auditorium at the megachurch in Singapore, I noticed rows of shelves in the gift shop advertising sermons previously delivered in the church and available for purchase. Many had titles addressing common emotional and psychological problems: 'Win over guilt and condemnation' or 'Turn your frustrations into breakthroughs'. Some sermons promised fortune and success or improved health and wellbeing. Others offered ways of dealing with conflict or reducing stress. New Creation Church is not just a place to receive pep talks and listen to inspiring lectures, however. Those who give themselves to Jesus receive in return a great variety of practical benefits: access to therapists, hospital visits, wedding venues, childcare, bereavement counselling, mother and baby groups, social support networks, and more. Selling points like these may help explain why charismatic Christianity has become the fastest-growing religion in Singapore and also why the brand has fast been going global.[2]

Although it may be tempting to dismiss the commercialized Christianity of New Creation Church as a cynical way of exploiting people's credulity, that would be to overlook the important social and psychological functions that evangelical religions fulfil. New Creation Church along with many other forms of evangelical religion provides for a wide range of basic human needs that are not easily met by other mechanisms or are more conveniently accessed in a religious setting. Although the goods and services on sale are mostly forms of psychological and social support that could in theory be procured from secular sources – counsellors, childminders, therapists, self-help books, social clubs, and so on – charismatic Christianity offers it all in a one-stop shop.[3]

New Creation Church raises revenue partly through a system of tithing, by which members can make regular payments. At one level, this is conceived as an expression of thanks to God for the various kinds of benefits conferred through the religious organization. As the church website puts it: 'Beloved, the Bible speaks of giving to God through our tithes and offerings. We believe giving is an act of worship and thanksgiving to the Lord, and our response to Him for all that He has done for us.'[4] At another level, it is analogous to paying upfront membership fees of a club, providing access to its various facilities.[5] Crucially, it also provides access to other members whose commitment to the beliefs and values of the church signal their reliability as cooperative partners.[6] Evangelical churches provide access to potential employees, trading partners, or job opportunities, and even act as dating agencies providing access to future spouses. The tithing system also allows prices to be adjusted to fit the pockets and incentives of different customers to maximize revenue. Richer members of churches operating this system are expected to pay more than poorer members. When people hit hard times, such churches can adjust their demands to fit the means of their flocks.[7]

Organizations like New Creation Church are no doubt attractive to members in part because they harness our religiosity biases. But they don't do so purely to line the pockets of charismatic leaders – they are also beneficial for their members. Like the moralizing religions that first appeared with the rise and spread of Axiality described in Chapter 5, contemporary forms of evangelical Christianity fulfil social functions that people value. Here, too, I am struck by resonances with the Kivung movement. It was seen by many of its critics as a cult, offering the false promise of supernatural rewards. But it provided a way of uniting a deprived and politically dominated portion of the country to fend off exploitative and destructive commercial interests and to provide much-needed infrastructure in rural areas. New Creation Church too is only in part about the promise of magical interventions; it also provides a means of addressing social problems in eminently practical ways.

The significance of New Creation Church, then, is not that it represents some radically new form of religion or social organization. In many respects, it is fulfilling precisely the function that fast-spreading churches have for millennia. Its significance is instead rooted in the way it appeals to our ancient religiosity biases using the latest innovations of the mass media and advertising. To the newcomer, there may be something distasteful, or at least disconcerting, about religion being dressed in the clothes of advertising, glitzy entertainment, and popular culture. But what if it were the other way around? What if the really shocking truth about today's mass media and the way products are sold to us online is that they have covertly assumed the status of our latest organized religions? What if they are hijacking our religiosity biases purely to make money in ways that are addictive and parasitic rather than to provide us with the kinds of social and psychological support we really need – still less to save our souls?

And what if they are now doing so in ways that are more Machiavellian than the methods adopted by Singapore's most popular megachurch, making them scarcely noticeable and therefore seemingly impossible to resist?

Animism and advertising

Commercial advertising plays upon the same kinds of biases as evangelical Christianity – even if this is far from obvious if you don't look closely. But unlike most organized religions in human history that helped bind societies together, the ad industry leads us into temptation and forms of mass consumerism that serve the interests of a tiny minority of unimaginably rich individuals at the cost of society as a whole. The high priests of this new form of religion are not popes, ayatollahs, or divine emperors but ad agencies.

Advertising starts to lay claim to our souls in early childhood. This first became apparent to me while studying how children become socialized into belief systems in which rituals and supernatural agents feature prominently. When I was applying for a grant from the European Research Council to study these processes, I emphasized the importance of research on the socialization of children into a world where organized religions continue to have a profound impact on the functioning of many societies and on geopolitical tensions around the world. But once I was awarded the funding, I became even more interested in the fact that we also fill our children's heads with a great variety of ideas about superheroes, human-animal chimeras, ghosts, witches, aliens, and a host of other supernatural beings. It is less serious in part because we think of these ideas as purely fictional. But these ideas also mould the ways we will later be exploited as adults.

Take Santa.[8] Children are invited to imagine this jovial bearded character in a red suit as a supernatural being, capable of visiting the homes of unfeasibly large numbers of children by means of the kinds of standard violations of intuitive physics we find in all religions. Santa offers up a dizzying panoply of minimally counter-intuitive ideas. The story involves the violation of intuitive gravity (flying reindeer), object coherence and solidity (associated with miraculous methods of ingress to children's houses), and mind-reading capacities (the ability to determine who is naughty or nice). These intuitive foundations might help explain why this tradition enthrals and delights vast numbers of people, young and old, every year; and why its commercial value is quite staggering: one 2016 survey valued the estimated holiday sales at over a trillion dollars,[9] more than the annual GDP of most countries.

If Christmas has provided a pretext for commercializing the intuitive tenets of both organized religion and folk tradition, it is not unique. Other public holidays arguably do much the same thing. Halloween, for example, may be seen as a way of commercially exploiting the same minimally counter-intuitive intuitions that prevail in wild religions everywhere. Like Santa, the stereotypical Halloween witch is often imagined as defying gravity (flying on a broom rather than a sleigh), as is the vampire (with a cloak that serves as bat wings). The counter-intuitive properties of zombies – another Halloween favourite – are particularly interesting from a psychological perspective. Here we are invited to imagine an agent with standard physics (no special ability to defy gravity, object solidity, or coherence expectations) but non-standard psychology, usually lacking free will and, in popular movies, normal cognitive functions (apparently being unable to speak, reason, or plan normally).[10] Instead, the zombie is endowed with rather unusual psychological traits, such as an addiction to cannibalism and especially the consumption of raw brains. In 2022, the National Retail Association estimated the value of Halloween in America to be worth over $10 billion. Similar forms of 'wild' religious belief can be found in Thanksgiving in the US (all that ancestor veneration) and Chinese New Year (a fire-breathing dragon being a neat case study of the appeals of violating intuitive biology). Thanksgiving costs Americans around $1 billion a year on the purchase of turkeys alone; in 2019, meanwhile, the value of sales associated with

Chinese New Year celebrations were estimated to be around $150 billion.[11]

Although the commercialization of religiosity bias is abundantly apparent on many public holidays, its benefits for consumers are far less obvious. Even if we recognize that we are being manipulated by seasonal advertising, we often assume that it is just harmless fun – an excuse for splashing out and spoiling the kids. But many of the needs we fulfil through these events in the ritual calendar are artificially created by profitable industries rather than being existing problems crying out for a solution.

The exploitation of wild religion does not end with us buying products that are merely useless, however. Advertisers are skilled at enticing us with wares that actively harm us and our environment (extreme examples being addictive commodities like tobacco). One of the commonest techniques used to ensnare us is to make products seem more like people. Just as wild religions everywhere anthropomorphize nature (seeing humanlike shapes and influences everywhere), advertisers do the same thing when marketing products. The anthropologist Stewart Guthrie has spent a long career exploring how the advertising industry utilizes the same anthropomorphic psychology as our most sacred works of religious iconography. He has identified myriad examples of ads featuring humanlike objects.[12] One shows coffee machines with buttons and knobs configured like faces discussing which coffee is best to use; another shows a telephone with a mortar board on top – the kind worn at graduation ceremonies – with an assurance that this appliance has passed more tests than a Rhodes scholar. Yet another shows a non-alcoholic beer aimed at a sporty clientele in which the branded beer bottle is wearing tennis whites. I have had the privilege of attending several of Guthrie's lectures in person and his slides displaying anthropomorphic advertisements have impressively diversified over the years. Observing such images in Guthrie's lectures juxtaposed with bas-reliefs, sculptures, wall paintings, and monumental architecture taken from religious traditions drawn from across the world through the millennia is an impactful experience.

Guthrie draws on the work of the sociologist Irving Goffman – who as early as the 1950s became interested in the way that themes of gender, sex, and fertility were used as 'bait' in both religious imagery and advertisements.[13] But while the baited hook delivers only disappointment for the unwary fish when swallowed, the sexually charged advertisement, like the decorated shrine, delivers a reward in the form of mild physiological arousal – albeit one that slowly reels us in. Over time, our desire for these rewards provides the basis for addiction. And, as Guthrie also observes, one of the common consequences of this form of addiction is the reinforcement of gender stereotypes that render women passive or subservient.[14]

Brands are increasingly able to quantify the effects of such anthropomorphic advertising on their commercial success. One review of over a hundred articles on the topic in 2022 concluded that portraying commercial brands as more humanlike can increase how much consumers like, are loyal to, and willing to pay a premium for products being marketed to them.[15] Interestingly, what seems to influence consumer preferences most is the extent to which commercial brands are conceptualized as agents with minds. In much the same way as our ancient ancestors attributed agency to animals, trees, and mountains, so too do we now attribute agency to brands.[16] We can track this phenomenon by inviting consumers to give ratings to organizations that reveal the extent to which they are perceived as humanlike. Studies in this area conclude that people regard brands like Nike, Microsoft, and Coca-Cola as relatively lacking in humanlike agency, being more or less comparable to trees, houseplants, and microbes respectively.[17] By contrast, brands like UNICEF or Teach for America are much more personlike, achieving scores close to those of an elephant or a corpse. Somewhere in between are brands like Facebook and Google whose scores place them on a par with an ant and a fish respectively, or, more positively still, a brand like Disney which comes close to achieving the same score as a bird or a mouse. By these measures, the world of brands appears very much akin to the animistic universe of the shaman. And just as cult leaders attract faithful followers, the evidence from marketing research indicates that brand anthropomorphism produces loyalty to favoured brands,[18] greater engagement with them via social media,[19] the cultivation of personalized relationships with favoured products,[20] and willingness to pay over the odds if necessary to preserve those relationships.[21]

One of the consequences of advertising based on the exploitation of our religiosity biases is that it can lead not only to positive anthropomorphic promotion of brands that ad agencies want to sell but also to the demonization of brands they wish to quell. Just as religious guilds and specialists often decry their rivals as deluded, subversive, and even satanic, similar strategies occur in the world of advertising. This phenomenon has been described as 'reverse brand anthropomorphism' and involves the vilification of brands, much as religious organizations have traditionally cast their competitors as evil, godless, or heretical.[22] A particularly striking example of this is the strange story of margarine.[23] Margarine was created in France during the Napoleonic wars in response to the need to supply the navy with a spreadable form of fat that didn't go rancid during long periods at sea. It never really caught on in France but later found a rich market in America, where it was used extensively in military catering in the Second World War and during post-war rationing. The dairy industry in America identified these developments as a serious threat to butter and instigated repeated campaigns to denounce margarine – not merely as a less tasty product but as a harmful and unnatural product, possibly linked to witchcraft. According to the nineteenth-century Republican congressman David Bremner Henderson, whereas butter was mentioned throughout the Bible, the first mention of margarine in history comes from the witches described in Shakespeare's *Macbeth*. In 1884 in Vermont and in 1891 in New Hampshire and West Virginia, thanks to lobbying by the dairy industry, new legislation required margarine to be dyed pink to make it less palatable.[24] This battle over what people should spread on their toast was at one level purely a matter of competition between opposing industries, but it was conducted in the same way that rival religions have traditionally fought over access to followers. Margarine was decried not merely as less palatable but as impure, contaminating, and evil.

The widespread use of anthropomorphic advertising is not merely a quirk of academic interest but constitutes a major threat to society at large – hijacking our religiosity biases to sell us things we don't need, and which harm the planet. Is there a solution? One answer lies buried in our collective past. Over the millennia, organized religions attempted to domesticate our propensities to believe in supernatural beings and forces by developing elaborated bodies of doctrine and practice that required extensive training and repetition to master. As part of this effort, many such religions – from puritanical varieties of Christianity to iconoclastic forms of Jainism – forbade or curtailed the use of anthropomorphism in religious art and architecture. This is why throughout much of the Muslim world even today, for example, decorative designs avoid representations of agency, particularly in its human or divine manifestations.

If we are to combat the scourge of advertising on our physical, mental, and societal health we might consider doing something similar. The mechanism adopted in organized religions has traditionally been to establish systems of formal education that provide systematic correctives to our intuitive theological misunderstandings and our penchant for the minimally counter-intuitive over the theologically correct.[25] And yet our secular systems of education today do little to correct the distorted realities that the advertising industry now transmits to each new impressionable generation. It is not compulsory at school to learn about the dangers of behaviour modification, or body dysmorphia, or the false promises of the cosmetics industry. Teachers are not obliged to tell our children about the benefits of recycling clothing, reducing wastage of food, or reducing our carbon footprints. Above all, our national curricula do not teach us about the psychological tricks and behaviour modification algorithms that advertisers use to sell us products, or how to manage them.

Even if public education reforms are slow in coming, faster solutions to the problems created by the excesses of advertisers might come from new technologies. A compelling example lies in the smartphone app known as Moralife: Global Consumer Databank. Moralife began as a database of foodstuffs but is gradually being extended to incorporate a wider range of popular consumer goods. It works like this. Each item in the database is evaluated according to a transparently quantifiable rating of morally high to low (i.e., good to bad). It works by tracking ever-changing factual information about the top fifty staple foodstuffs and the main multinational companies producing and supplying them. For each type of item, for example sliced bread, expert analysts rate leading brands according to various ethical dimensions such as how green they are (e.g., carbon footprint), how humane (e.g., RSPCA approved or not), how exploitative (e.g., salaries, working conditions) and for each of these domains they receive a score, updated following annual checks. These scores can be aggregated so that consumers know at any given time the overall ethical rating of a wholemeal baguette, for example. This information is collated by a special app suitable for most smartphones, tablets, and computers so that your weekly shop can be calibrated to a moral standard of your choice. Issues that matter more to you – let's say reducing carbon emissions or improving animal welfare – can be weighted to suit your ethical priorities. In other words, you can decide how best to calibrate your moral ledger as a consumer. As more consumers vote with their apps, more supermarkets and retailers will have to adapt accordingly – and the initiative is expected to spread to more and more companies and products. Unlike newspaper articles and online customer feedback – all of which can be hard to interpret or compare, Moralife allocates scores to a wide range of commodities based on consistent criteria, so you really can compare apples and pears.

How can you get the app? You can't, unfortunately. Moralife is just a figment of my imagination. But such things could be developed with relative ease. There are already apps that allow consumers to take a moral stand on products that fall short on ethical standards. For example, Buycott claims to provide shoppers with access to the guilty secrets behind product histories by scanning their barcodes – enabling customers to inform suppliers at the click of a button that they are boycotting their wares.[26] Similar ideas lie behind apps like DoneGood and Orange Harp.[27] If rolled out at scale, such apps could help to wrest control away from advertisers and into the hands of consumers.

However, apps of this kind would need to be adequately funded and supported if they are to become more widely adopted. Looking back over the grand sweep of human history, effective management and control of wild religion has generally required more than simply the moral commitment of the population at large. It has also tended to rely on the continuous efforts of centralized priesthoods and hierarchical systems of policing in order to maintain a stable orthodoxy. In secular societies, similar functions still need to be fulfilled; but this requires legislation and effective regulatory bodies. Manufacturers and importers could be legally obligated to provide ethically salient information needed for apps like Moralife to become officially certifiable and publicly available. We have already become accustomed to legislation requiring products to display health-related information. The suggestion that information on the ethical dimensions of products should be included simply takes that idea a little further.

Moral bias and the news

Just as wild religion has been commercialized through branding, the moralizing role of organized religions has now been increasingly taken over by the media, especially in highly secularized countries. Unlike organized religions, however, the media's role in establishing and enforcing trends in moral thinking is not geared to the needs of societies as a whole but only to minority vested interests. Whereas the moralizing religions that shaped the modern world managed to harness our moral intuitions in ways that served the interests of cooperation on ever-increasing scales, the manipulation of these same intuitions by news media is having a profoundly corrosive effect on society, reducing our confidence in political leaders and democratic institutions, and driving populism and polarization.

In so doing, the media too is appealing to our deepest intuitive biases. When we lived in small communities, in which everybody knew everybody else, news consisted mainly of socially strategic information about who was hoarding wealth, who was telling lies, who was sleeping with whom, who was stealing, who was free-riding, and so on. In most of these newsworthy stories, there would be transgressors and victims, and news purveyors and consumers would be very sensitive to the reputational consequences of this information. The common term for this is *gossip*.

Much of what we call 'news' in the contemporary world appeals to similar intuitions, even when it is presented as politics and current affairs. Following an analysis of 1,276 news articles in the UK press in a single month, researchers concluded that what leads journalists to decide whether a story is 'newsworthy' boils down to ten major considerations, which ideally coalesce to create a compelling story.[28] Does it concern powerful people or organizations? Are celebrities involved? Does it involve sex, drama, scandal, or other titillating or 'human interest' themes? Is it surprising? Is it upsetting (e.g., involving conflict or tragedy)? Is it uplifting (e.g., involving heroism, new cures)? Is it consequential? Is it relevant to people's concerns and interests? Does it connect with other recent storylines? Does it fit the agenda of the newspaper? Most of these are ingredients for interpersonal tittle-tattle as much as events of great public significance. In particular, a thread running through all these criteria is that a good story must incorporate morally salient topics. Who is lying? Who is acting selfishly? Who is cheating? Who is suffering? Who is being cowardly? Who are the bullies? Morally positive stories about heroes, innovators, rescuers, and saints clearly have a niche in this model but since bad events are more attention-grabbing than good ones, they generate more public interest and therefore more airtime.

There are evolutionary reasons why negative events are more newsworthy than positive ones. For example, overestimating risks can save lives and so is a cost worth paying. That doesn't mean we have no interest in good news, however. This too has its place in the top ten considerations of what makes a story pass muster with the average newspaper editor. But what all stories tend to have in common is that they are morally salient in some way and usually this is negative – pointing out transgressors and the grisly details of their crimes and the process of punishment.

In a local face-to-face community, the more socially damaging some talked-about transgression becomes, the more likely it is to escalate into something more than mere gossip and become a problem that would typically be addressed in a public setting. Communities everywhere have forums for this kind of thing, whether informal or legalistic, and what happens in those settings may be remembered or (in many literate societies) recorded more systematically and used as precedent when dealing with analogous situations in the future. The newsworthiness of all such events in a given community may at one time have stemmed from the fact that they have potential fitness consequences for others. Members of the community have a vested interest in ensuring that the way transgressions are dealt with is not systematically harmful to self or kin group. Not only can people remember who did what to whom based on direct observation but also, thanks to the evolution of language, based on second-hand reporting. And so eyewitness accounts, circumstantial evidence, and public opinion all inevitably become key features of norm enforcement and punishment. Effective reputation management now becomes vital for survival and reproduction.

If you were to randomly tune into the news anywhere in the world, you would therefore soon be sucked into narratives that are attention-grabbing because of their moral salience (and, consequently, capacity to generate profit for their purveyors). But whereas news of this kind would have been potentially very consequential for consumers in a small-scale society – in which everyone would personally know the key protagonists – in a centralized state any practical value diminishes. The moral transgressions of public figures you will never meet and whose personal lives do not affect you personally are interesting because we have an evolved predisposition to care about such things. Yet they are perhaps less significant than questions about whether these figures are managing competently the hugely complex bureaucratic institutions over which they preside or wisely investing public money in ways that are far from intuitively graspable.

The corrosive effects of this situation do not end with us being merely distracted by gossip, however. News media that continuously feeds us titillating and attention-grabbing stories is also corrosive to trust and societal cohesion. As we saw in Chapter 5, as societies grew larger, routinized forms of shared identity provided one of the main ways in which trust could be maintained at the scale of large communities. Eventually, the rise of moralizing religions provided ways of increasing confidence in the moral integrity of other people in your society, and particularly the holders of high office. As mechanisms for generating trust and legitimacy decline with ongoing secularization, the need for institutions fostering cooperation and trust grows ever more pressing. Unfortunately, the news industry is exacerbating rather than meeting this need.

Is there an alternative? Let us imagine for a moment if the news were to have an entirely different purpose – for example, to inform and educate citizens to make them more prosocial and critically alert contributors to the political systems in which they happen to live. If that were the explicit goal of news reporting, it would fundamentally change the way in which the news is produced and distributed. In short, it would mean rethinking what makes an event newsworthy.

Newsworthiness is not to be confused with debates about the quality of information. It is already widely accepted that many key organs of society must exercise tight control over the quality of information in the interests of accuracy and balance. Medical science, for example, seeks to ensure that the efficacy of a given treatment is tested via randomized controlled trials in large populations rather than decided by whose opinion gets the most clicks on news websites. Likewise, legal systems attempt to distinguish between hearsay and demonstrable fact. Something of the spirit of this approach even lurks in news reporting – for example, in the form of fact-checking mechanisms. But the question whether a given news item is accurate – or even whether it represents the opinions of people from a balanced selection of interest groups or political parties – must not be confused with the question of whether that item is newsworthy at all.

Indeed, the question of newsworthiness is, to my mind, the most fundamental but also the most underexplored theme in media criticism. Even the most laudable journalistic enterprises struggle to grapple with this problem. For example, the *New Internationalist*, founded in 1973, describes itself as beyond the influence of media moguls and corporate advertisers. It strives to cover topics that matter to humankind as a whole – going beyond the headlines in popular news outlets and exploring the underlying causes and wider ramifications of global events and trends. Nevertheless, what is missing from the magazine's self-presentation is a detailed account of how its journalists and editors determine newsworthiness.

If we the people are, by some democratic means, to be the ultimate arbiter of policymaking in the interests of humanity at large, then ideally the citizenry of the world would be knowledgeable about the ins and outs of cooperation problems facing all of us. Ideally, our intuitions about what is important to pay attention to in world events would be augmented and possibly even superseded by a more rigorous approach to the question of newsworthiness. We might see this as a process of domesticating wild religion, or in this case more specifically of domesticating 'wild morality'. Morality in the wild is all too easily exploited by commercial interests to drive news stories that make us more narrow-minded and adversarial. However, if the purpose of news media is to share information relevant to the health and management of societal problems, as a pillar of the democratic process, then what should make an event newsworthy is its impact on the functioning – for better or worse – of large-scale, complex societies. This would involve much more nuanced appeal to our moral intuitions, enabling us to weigh up conflicting impulses in light of the impacts of complex alternatives for public policy. Put more pithily, the challenge is to update the news by at least a few thousand years.

Such a process is unlikely to be led by the market, since the ultimate goal of journalism is to sell news and make profits. One way to tackle the ubiquity of the profit motive in news publishing might be to revivify and reform public service broadcasting. The main advantage of this approach is that it would enable news and current affairs programming to be funded by taxpayers and therefore to become relatively independent of vested commercial interests and the needs of advertisers. However, to reap the benefits of this independence, public broadcasting would need to be run in a way that prioritizes newsworthiness over ratings.

Candidate criteria for newsworthiness might include the following. First, does the news item identify a problem of sufficient scale to be a matter of public concern? In this respect, the apocryphal 'man bites dog' headline would be less newsworthy than new statistical evidence that dog attacks are spiralling of our control. All too often, public policy is driven by fears aroused by events so rare they amount to non-problems. In legal circles this is the maxim that hard cases make bad law (i.e., extreme cases are a poor basis for general laws). Second, does the event have significant ramifications? For example, does it have potentially enduring wider effects on some aspect of society or geopolitics and, if so, how? By making such effects a major focus, news would be transformed into a coherently unfolding narrative on changing political or economic landscapes rather than a process of piecemeal agenda-hopping. Third, is the problem identified fixable – and, if so, how? Focusing the news on solutions would increase the quality of public debate, especially if prioritizing structural problems and their practical remedies over individual transgressions and the punitive process. If one country, or a coalition of like-minded countries, were to spearhead a more principled approach to providing newsworthy news, this could serve as proof of concept, enabling governments around the world to borrow, modify, and build upon their methods to provide newsworthy news to their citizenry as a statutory obligation. To count as newsworthy, it would suffice for candidate events to meet any one of these criteria.

The creation of an informed global citizenry through the transmission of newsworthy news would probably never replace what we currently call 'the news'. There will still be ample opportunity for commercial media to sell advertising space through the spread of titillating gossip and divisive narratives. But an informed citizenry may increasingly see all that for the sideshow it really is. There is already some evidence that the more highly educated you are, the less time you are likely to spend watching television news broadcasts.[29] Perhaps, the more knowledgeable we become about truly newsworthy events the less we will be interested in the kind of news that currently dominates our media, or at least the less likely we will be to let it influence our more measured beliefs and actions as citizens.

Moral contagion and social media

Human cooperation depends upon third-party punishment: acts to sanction transgressions against persons other than yourself.[30] There is compelling evidence that even before we can speak, we expect people who behave immorally (e.g., who don't share things fairly) to be punished by third parties and for virtuous behaviours to be rewarded.[31] This principle lies at the core of all criminal justice systems, from the simplest forms of collective execution of dangerous psychopaths in a hunter-gatherer society, through to the most elaborate trials conducted in modern courts of law.[32]

Such behaviour is commonly described as altruistic punishment. The word 'altruism' is used in this context to indicate that there is a cost to the punisher that is not balanced out by gains. That is, the punitive action involved is not an act of retaliation but of public service, carried out in the interests of justice and protection for all. Of course, punishers do not really go unrewarded and so it may be somewhat misleading to describe them as altruistic. For example, the brain's reward structures are activated when miscreants are seen to get their comeuppance.[33] In fact, the neural architecture for punishment is highly specialized. Brain scans show that the prefrontal cortex is particularly involved in making judgements about the severity of punishments warranted for different types of crimes.[34] Using similar techniques, my colleagues and I found that activity in the same region of the brain was affected by levels of fusion measured in football fans when they were deciding whether to administer punishments to supporters of rival clubs.[35] Of course, punishers may also receive material benefits for meting out justice – most obviously in the case of salaries earned by police, lawyers, judges, jailers, and executioners. But even in cases where we are not actually paid to punish, we derive benefits from doing so in the form of positive reputational feedback. In all human societies, traditional or modern, people are admired for standing up to bullies, liars, cheats, and thieves. The root of that admiration is that we all know that this entails risk. It therefore requires courage.

The trouble is that social media platforms are now reducing these risks in unprecedented ways while nevertheless conferring reputational benefits virtually for free. Even though you may never have thought about it in terms of the costs and benefits of third-party punishment, the state of affairs I am describing will be familiar to anyone who has spent a meaningful amount of time on social media. In 2012, Mark Zuckerberg announced a 'tipping point' in our ability to use social media to share and consume information – and more specifically, our ability to endorse and spread news and comment we like or, just as easily, to 'unlike' people and retweet reputationally damaging information in an instant. In the years since, much ink has been spilled over the fact that we are suddenly living through a golden age of public retribution. Social media creates an environment in which people can shame, heckle, and denounce one another with remarkable ease; the creator of Twitter/X's retweet button, Chris Wetherell, once compared it to giving a loaded weapon to four-year-olds.[36] From an anthropological perspective, this is the age-old issue of third-party punishment (public shaming in this case) being used for selfish reasons (personal acclaim). Although this type of signalling evolved to facilitate cooperation in the group, such behaviours on social media may have less benign social consequences. The socially corrosive ramifications of this are complex.[37] But they point to a major dilemma arising from the transformation of the media under capitalism. It concerns the various ways in which our punitive instincts and acts of public shaming polarize our societies rather than bolstering cooperation.

The new era of public shaming brought on by social media is corrosive in part because it relies upon another of our deepest intuitions: the fear of contagion. Part of the reason why hanging out with a bad crowd is so risky is that the tendency to tar malfeasants with the same brush is deeply rooted in our ideas about contagion. In Chapter 2, I described not only how wild religion is rooted in panhuman intuitions about physical objects, mind–body dualism, and life after death, but also in various essentializing beliefs and fears of contagion. Online sharing mechanisms tap into fears of *moral* contagion. If an idea seems to have become widely tabooed, based on our perceptions of how attitudes are trending online, we might naturally dissociate ourselves from people who espouse those ideas and perhaps also from those who are publicly aligned with them. This fear of moral contagion is rooted in the same psychology that makes us reluctant to wear clothing that belonged to a murderer, as if the moral taint might be catching. To the extent that social media allows us to take sides at the click of a button, it encourages kneejerk tabooing of ideas that could be usefully discussed rather than being hastily cast out of the range of what is worthy of debate.

Imagine you are in a group that has tabooed a certain behaviour (let's say the use of pornography) and somebody in that group has been found guilty of engaging in that behaviour (i.e., viewing porn), then unless you feel a very strong sense of loyalty to the transgressor you may speedily dissociate yourself from the accused. The reasons why pornography use is proscribed in your community may not have much influence over your desire to distance yourself from it. You may disapprove of pornography on the grounds that it is exploitative, coercive, degrading, or addictive, or because of prudishness or religious piety. Or you may privately sympathize with the transgressor because you regard pornography as a legitimate industry or because you are a clandestine user of it. But in all these cases, the outward behaviour may be the same, which is to distance yourself from the miscreant. Social media plays into precisely this logic. It makes siding with the majority much cheaper, while increasing the costs of dissent. A study of Swedish Facebook users has shown that people are more than twice as likely to endorse a post by an unknown user of the platform (by clicking the 'like' button) if it has already been endorsed by three unknown users compared with only one.[38] Authors of the study concluded that this behaviour was motivated by a desire for reputational benefits via conformism (esteem, popularity, or respect among other users). These cheap signals cumulatively provide evidence of what 'everyone' believes, giving us all a superficial impression of consensus on moral issues and further deterring dissent.

The problems this poses for our public conversation are most acute when the judgemental culture of social media collides with issues on which there are genuinely deep divisions along moral lines. People's tendency to fear moral 'contagion' is nothing new – people have long sought to avoid or to shame those whose political opinions they find repellent, and this cannot be blamed on the internet or even on the retweet button. But what has changed is the scale on which we can take sides and the effects this can have on our collective conversation.

Take the case of Poland. In October 2020, Poland's constitutional court, the Constitutional Tribunal, ruled that abortion in the case of foetal abnormalities should be outlawed, triggering national protests on a scale unprecedented since the fall of the Berlin Wall.

Shortly after the protests, I teamed up with colleagues in Poland to design an in-depth survey investigating how moral intuitions informed both sides of the debate in a sample of over 500 Polish citizens.[39] We focused in particular on the ways in which our seven basic moral rules informed the thinking of pro- and anti-abortion supporters. As described in Chapter 2, we had previously identified seven forms of cooperation that societies around the world deem to be morally good: help your kin, be loyal to your group, reciprocate favours, be courageous, defer to superiors, share things fairly, and respect other people's property.[40] We now set out to examine how social media platforms play upon these universal intuitions – often using them, paradoxically, to further polarize debate.

What we uncovered was striking: that while everyone shares the same moral intuitions, they were being applied to the abortion issue in strikingly contrasting ways. Polish anti-abortionists were much more likely to prioritize deference to authority, caring for kin, and bravery – suggesting that they would have been especially moved by news stories that emphasized how the supporters of abortion showed disrespect (e.g., towards religion), poor commitment to family values (e.g., prioritizing selfish and individualistic goals), and cowardice (e.g., unwillingness to face up to the challenges and responsibilities of rearing a child, particularly one with congenital abnormalities). By contrast, we found that pro-choice supporters valued group loyalty higher than the opposing camp, suggesting that they were more likely to be moved by news stories emphasizing solidarity of a sisterhood or liberal ideological community. This would indicate that there are indeed some differences of opinion about morality, even if we all share the same basic intuitive assumptions.

Herein lies the problem with social media. The existence of both a broad consensus *and* genuine moral differences means that our moral intuitions are ripe for exploitation. If a site could ghettoize these differences of moral orientation within echo chambers, they could generate enormously high engagement – while also helping to harden political divisions and widen the gaps between opposing camps. This, it seems, is precisely what was happening in the Polish case. In the analysis of nearly a million tweets on Polish Twitter, researchers from the University of Pennsylvania found that influencers on both sides seldom communicated across the divide by engaging with each other's views and concerns.[41] Where there could once have been a debate that might build towards some kind of consensus, social media has created an environment that helps to harden pre-existing divisions.

What is to be done? A potentially effective way of enabling advocates of competing perspectives to work together to achieve more consensual and informed policy outcomes is to establish citizens' assemblies, which we encountered briefly in the last chapter. After many years of political deadlock over the issue of abortion in Ireland, a citizens' assembly of ninety-nine individuals was able to formulate a set of recommendations that enabled the people of Ireland as a whole to settle on a solution via a referendum, the outcome of which was to repeal laws effectively banning abortion. There seem to have been several reasons why this process led to such a clear consensus on a topic that had previously seemed insoluble.[42] One was that the assembly comprised citizens who had little opportunity to gain positions of power as a consequence of their deliberations. They were selected from the general population as a representative cross-section but with no hope or expectation of serving in a similar capacity ever again. Another key factor was that their discussions were extensively informed by expert opinion, allowing them to sift and debate research findings and their implications in a careful and methodical fashion. And yet another was that debate was conducted according to tightly enforced principles which included openness, respect, and collegiality.

What makes all these desirable features of debate and decision-making possible in a citizen's assembly but seemingly impossible to achieve on social media, is that the costs of third-party punishment are real and you have to face your critics and opponents in person. It therefore takes genuine courage to advocate a particular viewpoint because there is a risk of having the weaknesses of one's argument exposed and of losing support from one's backers as a result. Participants in the conversation therefore have to 'up their game' by improving the quality of their arguments and avoiding ill-considered rhetoric. Followers of competing viewpoints cannot so easily dismiss, 'unlike', or cancel ideas or people they find uncongenial. And as a result, the consensus tends to form around more moderate viewpoints, incorporating a cross-section of interrelated perspectives. Such a process naturally finds the points at which our shared underlying moral intuitions overlap, and it uses these as a foundation on which to build rather than to divide.

However, history shows that large-scale collective action problems are seldom solved only by individuals or even small groups. Although citizens' assemblies have a potentially important role to play in countering the corrosive effects of social media, we do also need effective governance mechanisms imposed from the top – because social networks seldom regulate themselves effectively at scale. This applies to social media platforms too. As early as 2011, arguments were being advanced for abolishing the retweet button. At the time, what concerned users was not so much the potential for hostile states to sow discord or for online bullies to victimize the vulnerable but an immediately noticeable deterioration in the quality of public debate.[43] Clearly, Twitter/X decided to stick with the button, however, and continued to do so long after it was being widely blamed for the rise of populism and polarization.

The extent to which regulation may be needed for different kinds of online activity varies widely. Low-risk forums such as hobby sites may require minimal regulation but those influencing election outcomes or driving forms of political extremism clearly do. The question, arguably, is not whether regulation is required but to what extent, where we source our models for regulation, and how we decide which models to implement. A first step might be something as simple as designing all platforms in a way that limits the number of times messages can be forwarded to larger groups. WhatsApp have introduced such policies from time to time and they seem to be effective in at least slowing the spread of fake news.[44]

Throughout world history, ever more sophisticated and elaborate governance arrangements have been necessary for cooperation to scale up. That process took thousands of years, as groups adopting successful strategies gradually outperformed those that didn't. But now that technologies are advancing at such a high speed, we cannot afford to rely on trial and error to winnow out harmful innovations. More urgently than ever, we need systems of governance that make use of much sturdier bridges between research-based policy recommendations and regulatory bodies – ones that take the most momentous policy decisions out of the hands of vested interests and place them firmly back into the hands of citizens working in partnership with trusted institutions of knowledge production, such as well-run universities and publicly owned broadcasting corporations.

When we see countries held in the vicelike grip of authoritarian theocratic governments, organized religion appears outdated, rooted in ancient principles of law, outmoded norms surrounding gender and sexuality, and theories of human origins that science has long since overturned. But if religion is here to stay – as I argued in Chapter 2 and 5 – then we surely need to have some way of harnessing and regulating this aspect of human nature. Throughout most of human history, organized religion provided institutional and ideological support for the cultivation of our moral priorities in society at large. With the separation of church and state during the European Enlightenment, responsibility for regulating wild religiosity became a private matter for religions and individuals rather than a function of government. This is less true of Islamic states and communist regimes. But now that so much of the world is organized at least in theory around principles of parliamentary democracy, unshackled by the dictates of any particular religious authorities, this has left a vacuum in the regulation of our wild religious predispositions and susceptibilities. That vacuum has been colonized by advertisers, news conglomerates, and social media platforms.

I have argued that this trend in the way we manage wild religion in the modern world has mostly been bad news for consumers, citizens, and societies. Organized religions once shaped and honed society's moral concerns over long histories of theological debate, continuously adapting to the changing needs of society. But today our moral sensibilities are being exploited simply to sell products, making tech platforms the new temples into which worshippers are being corralled. However, I have been arguing that the management of our religious and moral predispositions and susceptibilities would be better exercised by civil society than by commercial interests. The most effective way to do this is through a combination of grassroots democracy and top-down regulation. In this way, our natural religiosity can be managed by secular institutions for the sake of large-scale cooperation and not simply to line the pockets of ever-richer elites. If we want to introduce such regulation in fairer and more inclusive ways, then we will have to confront the corrosive effects of advertising, the media, and tech platforms more directly and decisively. In the next chapter, I argue that we also need to tackle yet another set of issues in the modern world that is rapidly reaching crisis point: the problems of tribalism, warfare, and exclusion.

9.
Tribalism Today

On 31 July 1994, I was making my way home along the Ormeau Road in Belfast, laden with shopping bags. My mind was reeling from an incident I had witnessed at the supermarket tills, where I saw a young mother repeatedly slapping her toddler for crying in the queue. Those small acts of violence may have been normal in the child's household, but it was preying on my mind. The impulse to intervene, to suggest an alternative way of responding to the child's distress, produced a stream of simulated scenarios in my head – all of which I realized would end badly. I was jolted from my thoughts by a series of gunshots, just yards away. Two men lay in a pool of blood on the pavement before me. The dead bodies belonged to two well-known loyalist paramilitary leaders: Joe Bratty and Raymond Elder. In the days and weeks that followed, the two instances of violence – the slap of the child and the killing of the two men – would soon become connected in my mind, the emotional impact of the one leaking into that of the other. It was almost as if they were aspects of the same event, perhaps the result of some elusive common cause.

What follows is partly an effort to find that common cause. From terrorism and warfare to gangland murders and football hooliganism, much of the violence we observe in the modern world is a result of three ancient forces: ingroup love, outgroup hatred, and norms condoning the use or threat of physical force. Understanding why this cocktail of ingredients is so volatile constitutes a necessary first step to preventing their most destructive effects.

Bratty and Elder were shot by members of the Irish Republican Army, who had for generations been fighting for a united Ireland between the predominantly Protestant north and the predominantly Catholic south. But by the summer of 1994, the winds of change were finally blowing and both sides in the conflict were on the brink of a ceasefire; one that would ultimately lead to an historic peace agreement in Northern Ireland. Weeks before the killings of Bratty and Elder, the Combined Loyalist Military Command (an umbrella organization representing a range of disparate paramilitary groups opposing a united Ireland) had seemed to be on the brink of calling a stop to the fighting. Some claimed that the IRA were taking out loyalist targets to settle old scores before any kind of peace negotiation could be entered into. Others thought the killings were tactical and future-minded, based on the logic that no long-term peace would be possible when violent extremists like Bratty and Elder continued to occupy positions of leadership.

In Chapters 3 and 6, I argued that identity fusion can motivate willingness to fight and die when the ingroup is threatened. This might explain why the IRA were willing to carry out the killings on the Ormeau Road even though the risks of getting killed themselves, either by armed loyalists or by members of the Royal Ulster Constabulary, were dangerously high. When the IRA operatives ditched their van and ran for cover, they did so with the support of parts of their local community, who effectively shielded them from the RUC while they made good their escape. Many members of that community were probably fused as well.

Indeed, the longer I lived in Northern Ireland the more I suspected that much of the motivation to fight and die in the war between nationalists and loyalists was rooted not only in hatred but in love – especially love of family and community. I was often struck by the similarities between warring tribes in Ulster and in regions of Papua New Guinea where histories of feuding and violence were rife. Not only was the scale of many of the attacks and counter-attacks comparable, with those on both sides knowing many of their enemies by name and reputation, but the leaders of various paramilitary organizations in Northern Ireland also reminded me of so-called 'big men' in the New Guinea Highlands. Both fought not only on the battlefield, but also in highly ritualistic battles – whether they took the form of Melanesian feasts or the marches of Orangemen. Above all, both these forms of intergroup hostility and violence were deeply rooted in history. As far back as social memory can take us in New Guinea and Northern Ireland, embattled local groups have settled grievances in tit-for-tat violence, raiding, and guerrilla-style warfare based on visceral feelings of shared suffering and intense loyalty to group and the thirst for revenge that this inevitably produces.

In those days, I didn't have the words to describe this process in detail. But looking back now, I suspect that I was witnessing the same processes I would later study in many other conflicts around the world – often on a much larger scale. Identity fusion – which originally evolved to support resilience and cooperation in small groups – now unites much larger groups than ever before in human history, largely due to the rise of media reporting. These days, news stories can plausibly spread feelings of shared experience at the level of transnational diasporas and world religions. Until recently, the sharing of intense and personally transformative experiences was limited to direct, in-person experiences of events like state-sponsored ceremonies or large-scale massacres and invasions. But with the development of more sophisticated forms of communication and media coverage, intense experiences can be shared vicariously almost instantaneously – everywhere from news camera footage to videos on social media. This provides a radically new mechanism for shared experience and extended fusion across much larger populations.

As a result, support for military action can be generated very quickly across vast territories, in some cases driving both states and non-state armed groups into violent forms of intergroup conflict. And this is no longer a competition any of us can win. As we saw in Chapter 6, the ability to unify populations behind hawkish policies was once both a curse and a blessing: for all its horrors, violent conflict served as a primary mechanism for growing social complexity. Today, the opposite is the case. The increasingly destructive power of today's weaponry means that warfare driven by tribal instincts writ large can only lead humanity down a path of self-destruction.

The problems posed by these newly expanded forms of tribalism are not only to be seen in how we fight our enemies but also in how we mete out punishments within the group. For most of human history, the idea of locking up large portions of the population as a way of preventing violence was unheard of. Yet in the last two centuries, it has become standard practice in every country on the planet. Most liberal democracies today incarcerate and exclude large sectors of the 'tribe', stigmatizing them as members of outgroups and impeding them from readmission, even though the costs of this to society are severe and the benefits obscure. What drives this socially dysfunctional mode of punishment is the same type of tribalism that motivates all forms of outgroup hatred. But stigmatization of criminals is justified not on the basis of class, caste, or race (at least not necessarily in an overt or explicit way) but in moral terms.

I will argue in this chapter that the antidote to these dysfunctions is not less tribalism, but more. If the mechanisms of fusion and identification we have inherited from our forebears can drive us into wars and other intergroup conflicts, they can also lift us out of them. Rather than thinking of tribalism simply as a problem we need to solve by trying to tamp it down or eliminate it, it makes more sense to think of it as a resource that can be used for good or ill. Tribalism can lead us to lock up vast portions of our citizenry, but also be used to bring them back into the fold. Tribalism can escalate warfare, but it can also be used tactically to produce the opposite effect.

Decades later, I would come to believe that this fact explains the missing link between those two acts of violence: the child being slapped in the supermarket, and the IRA hit on Bratty and Elder. Violence is not some inevitable consequence of tribalism. Rather, it is often used strategically, even reluctantly, when tribalism has gone wrong. Perhaps the IRA believed that their localized, one-off attack would help to end larger-scale cycles of violence in the long run; perhaps the mother in the queue believed that hitting her child would help successfully socialize him into a world in which violence was normal. This is not to say that either party was correct in their analysis or justified in their actions – merely to acknowledge that the link between identity fusion on the one hand and violent forms of pro-group action on the other is not automatic.

Fusion only turns to violence when peaceful alternatives are considered lacking or less effective, as well as when violence is at least in theory normatively acceptable as a way of handling conflict. Understanding this process – and finding ways to intervene – is the first step to reducing violence around the world.

Fusion, the media, and violent extremism

Around ten years after leaving Northern Ireland, I would find myself in a humid office in the heart of Jakarta seated opposite Nasir Abbas. Once a senior member of the Indonesian terrorist organization Jemaah Islamiyah, Abbas received his military training in Afghanistan and later helped set up a training camp for jihadis in the Philippines. I had come to talk with him about his work with convicted terrorists in Indonesian prisons – carried out with the assistance of the Indonesian police and security services at the highest level.

Why, you might ask, would convicted terrorists want to work with him – let alone Indonesia's law enforcement and anti-terrorist units? These questions were whirling around in the back of my mind as I tried to remain focused on the purpose of the meeting, which was to learn how support for violent extremism can be fuelled by portrayals of violence and injustice in the media. Nasir Abbas and his colleagues from Persada Indonesia University had been developing a technique for exploring how convicted terrorists started on the pathway to radicalization. They had persuaded many prisoners to tell them their life histories, using the metaphor of a river. Each participant in the study traced his life course on a sheet of paper, starting with their birth at the river's source and then recording key life moments along the course of the river as it wended its way slowly down through the landscape. But whereas on a normal river such landmarks might have been a village, farmhouse, paddy field, or stretch of wilderness, these points on the riverbank depicted dark nights of the soul linked to experiences of outrage, moral self-searching, violence, arrest, and incarceration.

I was continually distracted in our discussions by my curiosity about Abbas himself. He was affable, intelligent, and softly spoken, with an occasional flash of wit as he alluded to the ironies of loving God while waging war. I found it hard to imagine him planning terrorist attacks. I wanted to talk about the river flowing through his own life, starting with his experiences of radicalization at school, the impact of scenes of atrocities committed against fellow Muslims on the television, how he persuaded his father to let him go to Afghanistan to receive military training, how he spent his periods of leave on the frontline, learning how to use small arms and operate guided missiles. I longed to hear about his work in the Philippines, how he became embroiled in Bin Laden's network of allies and operatives, and how he felt about the Bali bombings and other atrocities. How did he end up being arrested in Indonesia and daring his armed guards to kill him, throwing punches and kicks to taunt them into opening fire? In a previously published interview, Abbas described how he had reflected long and hard on the question of why he hadn't been killed by the guards – who, after all, had been trained to shoot when provoked. Indeed, this proved to be one of the key landmarks on the river of his own life. He had concluded that God's purpose was for him to change his course and help the Indonesian government's deradicalization efforts instead of returning to his work at jihadist training camps.[1]

This was quite a turnaround. The pathway to violent extremism had begun over thirty years previously, when Abbas was a teenager. He read newspaper articles vividly describing the sufferings of fellow Muslims in Afghanistan at the hands of Russian troops – an experience that threw him into a current that would ultimately draw him to terrorism. Although Abbas remembers the impact these newspaper stories had on him as a teenager in the early 1980s, this was already becoming an outmoded pathway to radicalization. Moving pictures were rapidly taking over from the written word as a way of sharing experiences with the victims of faraway atrocities reported in the news. According to research carried out by the Profiles of Individual Radicalization in the US (PIRUS) project, the number of violent extremists of all stripes whose journey on the path to radicalization began with television news reports increased dramatically in the second half of the twentieth century.[2] The rise of social media in the twenty-first century strikingly exacerbated the problem: between 2005 and 2010 only 27 per cent of people radicalized in the US were influenced by social media compared with 73 per cent between 2011 and 2016.[3]

We saw in Chapter 6 how extended fusion in large groups often derives from shared emotionally intense experiences, which are also integral to people's personal identities: the 'imagistic pathway' to fusion. New forms of media have made this possible on an unprecedented scale. Today, we live in a world in which acts of violence against religious and ethnic groups in regions like the Balkans and the Middle East are experienced vicariously online or via television screens. Images of violence are broadcast not only through news reports from journalists on the frontlines but via social media uploads from smartphones and video cameras of ordinary people living in war zones all around the world. A striking example of this is the way events during the Arab Spring were shared by millions of people in Southeast Asia, unifying Muslims across the world around experiences of suffering shared through mobile devices and televisions.

Evidence from our surveys in Indonesia suggests the vicarious sharing of these experiences can have a profound impact on large-scale group cohesion. In one of our first studies on this topic, my colleagues and I gathered data from 1,320 Muslims in Jakarta, to explore how vicariously shared experiences among Indonesia's

Muslims gave rise to extended fusion with vast numbers of other people.[4] In particular, we wanted to compare these processes across various religious groups on a spectrum from moderate and peaceful to hardline and hawkish. For this purpose, we focused data collection on three sectors of the population: Muslims who attended mosque but were not associated with any kind of activist organization; members of Nahdlatul Ulama (NU), a moderate Muslim organization; and members of the Prosperity and Justice Party (PKS), a more hardline fundamentalist group affiliated to the Muslim Brotherhood in Egypt and other transnational Islamist organizations. We asked participants to write about experiences that they felt were most defining both for themselves personally and for the Muslim groups they belonged to, including 'all other Muslims in Indonesia'. Since the people surveyed were free to choose which experiences to focus on, there was quite a lot of diversity of choices among respondents. Nonetheless, we found clear differences among the various categories of participant. Members of the public at large were most likely to focus on the conviction of Jakarta's governor in 2017 for the crime of blasphemy, after he was accused of insulting Islam by referring to a verse in the Koran in one of his campaign speeches. For many moderates in the general population, the two-year prison sentence imposed on the governor, Ahok, was unduly harsh and many poured onto the streets protesting against the ruling. By contrast, hardliners in Indonesia also turned out to protest, proclaiming that the punishment was too lenient. Polarization over this issue was fuelled by media reporting that emphasized themes of shared suffering, occasioned by religious intolerance and blasphemy laws in the case of the general population and insults to Islam in the case of hardliners.

The extent to which experiences of suffering were indeed felt to be both shared and transformative predicted levels of fusion with extended groups of fellow Indonesians, in much the same way as directly shared experiences drive local fusion in relational groups. But by focusing on experiences vicariously shared on a large scale, we were able to see how events reported in the media could have life-changing impacts on people – not only on their personal identities but also on the way this affected their relationships with vast imagined communities. If you belong to a group that espouses hardline doctrines and policies, such as the PKS, extremist beliefs become your sacred values – whereas, if you belong to a moderate group such as NU or the general population, more tolerant beliefs become your sacred values. Support for collective action is stoked by the media in line with the norms and values of the groups that matter to us.

As reports from war zones become increasingly immersive, they record not only the devastating effects of bombs and bullets on the ground but also the sufferings of children in hospitals and elderly people weeping on camera, all narrating their experiences of grief and loss. News feeds are no longer simply informing us about world events, they are encouraging us to experience them in increasingly visceral ways. Televisions, computer monitors, and smartphone screens are blurring the line between lived experience and narrative, and between relational groups and imagined communities. And the ability of the media to fuel extended fusion and the will to fight has grown far stronger as reporting of the horrors of warfare becomes ever more up-close and personal.

It is not only conventional media that are increasingly contributing to the power of extended fusion to mobilize conflict efficiently and at high speed around the world. We saw this in many countries caught up in the Arab Spring in 2011. As is well known, the chain reaction of revolutionary movements across North Africa and the Middle East began when footage of street protests caught on smartphones went viral on the internet following the suicide of 26-year-old street vendor Mohamed Bouazizi in Tunisia. In richer countries like Libya, citizens acting as amateur journalists uploaded images of the unfolding conflicts on social media platforms like Facebook, YouTube, and Twitter as they unfolded – likely driving extended fusion among the revolutionaries we met in Chapter 6. As dramatic events were spread via smartphones and tablets, they fuelled mass unification and protest across the region. Since then, tech platforms have only grown as a key tool in the mass mobilization of populations behind political goals.

A clear example of this is the use made of Twitter by militant Islamic fundamentalists in Afghanistan. In April 2021, following NATO's withdrawal announcement, the Taliban launched an offensive against the US-backed Afghan government. The insurgents wanted to create a new Islamic state implementing their version of sharia law but on paper their chances of success looked very poor. They lacked fighter jets, heavy artillery, or professional military training and they were vastly outnumbered by the Afghan army. Yet they succeeded in overthrowing the government and taking the country by force through wave after wave of successful offensives in which the opposition seemed to dissolve before them. How was this possible? According to a 2022 report by a consortium of conflict researchers, a major factor in the Taliban's success was their ability to weaponize Twitter.[5] The report, co-authored by Brian McQuinn – the colleague with whom I had visited Misrata in 2011 – showed that the leaders and supporters of the Taliban used Twitter to create potent feelings of shared suffering and moral outrage in the general population of Afghanistan and to encourage members of the official army to defect. Researchers analysed activity on sixty-three accounts claimed by the Taliban which, according to the report, generated over two million followers using the retweet button to spread anti-government propaganda. These tweets included graphic images of atrocities attributed to US drone attacks, evidence of mass defections on the part of the Afghan army, and news reports of other alleged war crimes against the Afghan people. The report argued that the use of Twitter by the Taliban played a crucial role in the takeover, strengthening support for the rebels in ways that played upon images of shared suffering and contributed to the collapse in support for the government.

The vicarious sharing of transformative experiences via the media, and the forms of extended fusion this creates, will only increase in scope and intensity as communications technologies create ever more immersive reporting. We may wish that news journalism would focus more on the key geopolitical implications of conflict and the strategies for peaceful solutions, rather than on the human stories that drive human passions and bloodlust, but the latter is what sells stories.

But this does not mean the situation is hopeless. There are two reasons for optimism. The first is that we have increasingly effective methods for establishing when and where extended fusion is likely to produce positive effects on peaceful cooperation versus destructive and deadly outcomes. The second is that extended fusion alone is not the problem – in fact, there are many positive consequences of being able to share experiences on larger and larger scales. The real problem is when extended fusion becomes combined with perceptions of outgroup threat and violence-condoning norms. How might each of these opportunities be utilized to our collective benefit?

Detecting violent extremism before the terrorists act is not easy, even with a deep understanding of ingroup psychology. For every would-be violent extremist there may be countless individuals who express extreme beliefs but do not themselves pose a danger. Law enforcement and deradicalization programmes may therefore be missing the signs that really matter. As a result, they may end up pursuing the wrong people – those who give vent to extreme forms of outgroup hatred but are not really a security threat. In some countries, including the UK, making public statements encouraging support for prohibited groups is a crime carrying heavy custodial sentences, but it is questionable whether this approach reduces terrorism risk. It may have the opposite effect of exacerbating perceptions of outgroup threat, oppression, or persecution.

To help develop more effective diagnostic tools, I teamed up with Julia Ebner, an expert on undercover research with extreme groups, and Chris Kavanagh, a stats whizz who had worked with me for several years to design and implement various projects using fusion measures. The challenge was to find a new way of spotting high-risk individuals in the crowd, based on clues given away without their being aware of it. We were not seeking to provide evidence on which to base arrests but only to establish signals indicating risk, so that effective interventions could be attempted before it was too late.

In our first study, we set out to analyse the language used in fifteen manifestos written by a diverse sample of activists.[6] At one end of the spectrum were convicted terrorists who went on to carry out deadly attacks in Europe and the US and the leaders of violent right-wing and Islamist organizations. On the moderate end we included manifestos of reformers advocating various forms of social justice or action on the climate crisis. In the middle we included the manifestos of individuals espousing ideologically extreme ideas on both the left and right but who had not engaged in acts of violent self-sacrifice.

Our goal was to look for language in these manifestos indicating the lethal combination that, as we saw in Chapter 3, drives violent extremism: fusion, outgroup threat, norms condoning violence, and dehumanising beliefs. In order to single out the highly fused writers of manifestos, we needed a way of spotting tell-tale signs of this in their writings. We chose to focus on language referring to ties of kinship and shared biology as a reliable proxy. This was based on the growing body of evidence we had gathered showing that when people become fused, they perceive others in the group as like family and spontaneously use the language of kinship to express it.[7] We meanwhile coded as evidence of perceived outgroup threat all instances of language indicating that a hated enemy was acting in ways that harmed the ingroup. Twenty-four students were recruited to analyse the language used in the manifestos. We made sure they were unaware of our predictions, so as to minimize any effects of bias in interpreting the signs of fusion and threat. What this revealed was that 89 per cent of the manifestos displaying the tell-tale combination of deadly predictors were written by individuals who went on to commit acts of violent extremism. By contrast, individuals who wrote ideologically extreme manifestos but who were non-violent did not exhibit this combination of attributes.

Of course, detecting potential killers in the crowd is only one aspect of the challenge faced by counter-terrorism efforts. There is also the question of how to prevent people from becoming violent extremists in the first place. If the fusion-plus-threat model is central to that process, then interventions need to address the roots of fusion itself. Our research suggests that the two main pathways to fusion are perceptions of shared experience and feelings of kinship. Unlike political and religious convictions which people may disguise for fear of persecution or arrest, personal experiences and family ties are topics on which people are often willing to reflect freely and talk at length. The connection between these topics and the will to go to war is not always obvious. But that is also why they may present an ideal starting point for tackling outgroup violence.

In theory, if people are willing to talk about the life experiences and relationships that underlie their commitment to the group, then it is also possible to reframe their ideas about the sharing of group essence, which lie at the heart of many forms of violent extremism. A central idea here is that when people gather regularly to discuss their experiences and talk through areas of conflict, they may discover that the experiences of ingroup members are more diverse than they imagined, while commonalities of experience *across* groups are more similar than previously considered.

One reason why we generally overestimate the extent to which our own thoughts and feelings are shared with the group is due to so-called 'false consensus bias' – the tendency to assume that whatever you think and do is what others would also think and do.[8] For example, if you hold a negative view of people in another group, you would likely assume the majority of your fellow group members share those attitudes; whereas if you hold positive views of an outgroup you would likely attribute the same opinions to most other people in your group. However, if you were to listen closely to what other members have to say on the topic of group-defining essences, you might discover some surprising ingroup divergences as well as outgroup overlaps, reframing and nuancing your relationships both within and between communities. Fostering dialogues in this way may dilute the strength of parochialism and xenophobia – reducing tendencies towards violent extremism on the periphery of the group but also leading to more moderate attitudes at its centre.

This can often happen naturally in sectarian conflicts as soon as peace agreements take effect, if only because the reduced risk of being punished for speaking out emboldens people to share their experiences more publicly. In Northern Ireland in the mid-1990s, this was particularly noticeable in comedy clubs, with stand-ups now saying the previously unsayable about experiences of shared sufferings on both sides of the conflict. My colleagues and I also recorded comparably positive effects on fusion across sectarian divisions in Bangsomoro after it gained autonomy from the Philippines in 2019 following a history of intense violence and protracted peace talks. Once the weapons were laid down, we discovered that the same feelings of fusion which had previously driven violent extremism could become a force for reconciliation.[9]

Tribalism in the age of mutually assured destruction

Throughout recorded history, the success of the world's great armies has depended on high levels of military cohesion: from the armies of ancient Greece to the jihadis and crusaders of the Middle East. Contemporary psychological research has shown that fusion binds together fighting forces like no other form of social glue. But today, this poses an existential threat. If the most destructive weapons at our disposal were still swords, longbows, cannonballs, or even bullets and conventional bombs, the more cohesive armies of the world – especially those fuelled by radicalism and revolutionary ideologies – might well eventually prevail over less passionately committed adversaries. Yet we have now reached a threshold in the evolution of military technology at which extended fusion within militaries raises the prospect of mutually assured destruction due to the proliferation of nuclear weapons and the spectre of ever-deadlier forms of biological warfare.

The problem originates in deep-rooted features of our collective psychology. Living in large groups makes us prone to simplistic forms of us–them thinking. When we align with imagined communities based on ethnicity or doctrinal religion we increasingly see one another as cardboard cut-outs rather than rounded individuals with unique qualities and personal histories. Pigeonholing other people in this way improves the efficiency of certain kinds of interactions between strangers, but it can also lead to overly simplistic stereotyping and us–them thinking. To make matters worse, identification and extended fusion with group ideologies can exacerbate various forms of bigotry associated with negative stereotyping. For example, when we identify strongly with an ideological position, it becomes harder to countenance contrary viewpoints. And as our research with religious fundamentalists in Indonesia suggests, extended fusion can further harden ideological commitment, increasing our resistance to argument-based or evidence-based rebuttal.

It may seem that this problem is inescapable. In a big society, where anonymous interactions are frequent, it is often quite efficient to adopt simplistic stereotypes in ephemeral interactions. Stereotypes serve as rules of thumb for making inferences about people in order to help us predict their behaviour and adjust our own. Relying heavily on such rules of thumb would make no sense in small groups because we have far more detailed and nuanced information about other group members and can make more accurate predictions about their likely reactions in different settings than any simplistic stereotype would allow. This may help to explain why people find it hard to square their stereotypes with personal experiences. For example, when someone makes a racist remark, they might feel a sudden urge qualify it by saying, 'Of course, not so-and-so' (referring to some friend or neighbour from the ethnic group in question whom they happen to know quite well, and who 'obviously' isn't like that). Stereotypes seem most readily applicable only to 'the rest of them' – those others who are more typical of the group they come from but about whom we, seemingly by coincidence, happen to know very little. The more time and effort we invest in getting to know someone, the richer our knowledge of that individual becomes, and the less adequate stereotypes seem.

If we are prone to thinking too simplistically about people who aren't like us, then what happens when this is scaled up to the level of international relations? Not long after the Vietnam War ended, researchers developed a construct called 'integrative complexity' – or 'IC' for short.[10] The abbreviation was pronounced 'I see', neatly conveying the idea of being able to see other people's points of view and to recognize the multifaceted nature of the social world around us.

Integrative complexity is a measure of how capable people are of integrating multiple perspectives and their ramifications when reasoning about a particular problem. Consider, for example, the differences between two clauses in the Treaty of Versailles.[11] The first (Article 231) read as follows:

> The Allied and Associated Governments affirm, and Germany accepts, the responsibility of Germany and her allies for causing all the loss and damage to which the Allied and Associated Governments and their nationals have been subjected as a consequence of a war imposed upon them by the aggression of Germany and her allies.

However, as a result of considerable back-and-forth between diplomats representing both Germany and the Allied Nations, the second (Article 233) recognized many more potential contingencies that would need to be taken into consideration:

> The Commission shall consider the claims and give to the German Government a just opportunity to be heard . . . The Commission shall concurrently draw up a schedule of payments prescribing the time and manner for securing and charging the entire obligation within a period of thirty years from May 1, 1921. If, however, within the period mentioned, Germany fails to discharge her obligations,

> any balance remaining unpaid may, within the discretion of the Commission, be postponed for settlement in subsequent years, or may be handled otherwise in such manner as the Allied and Associated Governments, acting in accordance with the procedure laid down in this part of the present Treaty, shall determine.

Whereas the second Article considers a range of possible outcomes and how they should be handled (high IC), the first one is little more than a simplistic allocation of blame (low IC). It is important to emphasize that low IC does not mean 'bad' and high IC 'good'. In an emergency you might want communications to utilize low IC because appraisals and action plans may need to be simple and implementation unquestioning. However, when there is time to reflect on the ramifications of a course of action that involves many parties with differing goals and perspectives, high IC often delivers much better outcomes.

Researchers have found that when governments are building up to launching an attack on another country, IC levels in their communications go down. By contrast, communications from the countries coming under attack tend to be higher, only dropping to the same levels as the aggressor after warfare has broken out.[12] The same has been found of extremists: studies of the statements made by terrorist groups suggest that levels of IC drop significantly as attacks become more imminent, suggesting that the ability to process the complexity of intergroup conflicts reduces as commitment to violence grows stronger.

Low IC may contribute to intergroup conflict because dogmatic rigidity and unidimensional thinking, as well as difficulty integrating diverse sources of potentially relevant information, make it harder to identify win–win outcomes in adversarial situations. For example, if you are quick to issue ultimatums without considering whether your enemy has the capacity to comply, you may eliminate more promising avenues to conflict resolution. Low IC may also lead us to blame other people's failings on inherent deficiencies rather than potentially resolvable problems for which they aren't responsible, at the same time as it prompts unwarranted confidence in one's own opinions and judgements. Such traits tend to result in unproductive blaming and self-righteousness, acting as a bar to productive negotiation and compromise.

To the extent that low IC contributes to dogmatism and outgroup hostility it may make it harder for us to navigate a safe passage through the nuclear age. Reassuringly, however, the converse is also true. There is evidence that higher levels of IC lead to more cooperative solutions in negotiations between adversaries.[13] And there is evidence that IC can be increased with relative ease. It would be a mistake to think of high or low IC as equivalent to personality traits or IQ levels. The latter are relatively stable features of our individual psychological makeup that we carry with us through life into the many different contexts that we encounter. But IC is highly context-dependent and therefore potentially malleable. I may show an extraordinarily high capacity for integrating complex ideas in one domain – say, helping a friend to manage the challenges of a major setback such as a failed relationship or loss of earnings – but suddenly become dogmatic and inflexible on certain topics where I have firm opinions, particularly when tribal thinking comes into play. Even our most intelligent and seasoned leaders may become surprisingly rigid in their thinking when it comes to intergroup conflict and warfare.

For example, it is hard to deny that Tony Blair was an extremely successful UK prime minister, who would surely have scored high on IC measures across a very wide range of topics during most of his premiership. And yet when it came to the question of the threat posed by Saddam Hussein in Iraq, Blair was prepared to risk everything, including his own career, to mount an invasion in concert with US forces. Whereas a decision so momentous would normally have required exceptionally robust intelligence on the ground, Blair's decision seems to have been swayed too readily by circumstantial evidence of Saddam's intent to use weapons of mass destruction. It turned out after the 2003 invasion that there were no such weapons in Iraq. Nor did the invading allied forces have adequate plans for establishing a peaceful system of government in the aftermath. Gut feelings appear to have swamped all other considerations.

Political leaders do not have to fall into such traps but can take steps to overcome any temptation to indulge impulses driven by low IC. In practical terms, this might mean overriding the will to war, when fusion and threat conspire with military capability to motivate invasions and military interventions. When tribal instincts are at their strongest, this is when we need to lean more heavily than ever on evidence rather than hunches, and on rigorous modelling rather than propaganda. We need robust and defensible principles to guide foreign policy, such as the pursuit of minimum interference for maximum humanitarian benefit. We need interventions to be future-oriented and not focused exclusively on short-term alleviation of symptoms. We need to find ways of fostering realism and managing overconfidence in warfare as a solution to international disagreements.[14]

The practical implications of this kind of research for addressing problems of conflict prediction and prevention are particularly promising. Research into the role of IC in international conflicts has already demonstrated its relevance to areas of international relations associated with forecasting, intelligence, and diplomacy.[15] To the extent that deterioration in IC is measurable and its implications for the risk of warfare quite well established, it could be used more systematically by foreign affairs experts and diplomats to adjust communications strategies and to head off potential breakdowns in peaceful negotiations.

A similar role could be envisaged for the type of natural language-processing analysis that my colleagues and I have developed in efforts to predict and prevent acts of terrorism.[16] As I explained early in this chapter, our methods could be used to pick out those most likely to undertake acts of violent extremism from a much larger pool of ideological extremists, but the same approach could help to identify which authoritarian leaders will be next to surprise the world by launching attacks or invasions that, from a more hard-nosed practical perspective, seem foolhardy or irrational. At my college in Oxford, Julia Ebner and I have established a project investigating the psychological foundations of institutionalized extremism in autocratic states entitled 'When Despots Become Deadly' aimed at supporting early conflict-prevention efforts in UK foreign policy.[17]

Beyond the level of individual states, an obvious forum to develop a more thoroughgoing and rigorous framework for such approaches would be the United Nations. Building on its existing role in international peacekeeping and preventative diplomacy, the UN could do more to harness the explanatory power of the new sciences of tribalism – especially fusion theory and research on IC – to develop early interventions before conflicts escalate rather than being primarily reactive in the wake of widespread bloodshed. Establishing a more effective forecasting and conflict-management unit within the UN would be relatively easy but for it to have the power to overcome more parochial forms of tribalism and the lowered levels of IC that these produce, then the UN would need to become a much stronger force in global affairs than it is currently.

Scaled up, such approaches would amount to a more rigorous and empirical approach to foreign policy than ever before. But at the same time, they would be reviving the time-honoured role of organized religion discussed in early chapters – only in the secular sphere of international relations rather than the realm of spiritual or theistic beliefs. The doctrines of this new sphere may be grounded in secular theories of cohesion rather than religious or theological ideas but they would be no less psychologically compelling and socially impactful. In offering a systematic departure from 'wild' or intuitive thinking and replacing it with a more coherent and consistent reliable body of thought, these new approaches would harness the predictive power of scientific methods, in place of faith-based appeals to the power of prayer.

Returning to the tribe

In early 2023, I met the Vanuatuan politician Andrew Solomon Napuat at a waterfront bar in the country's capital, Port Vila. As I waited for him to arrive, I gazed out over the vast expanse of the ocean through the fronds of palm trees but found myself thinking about the prisons I had just driven past with claustrophobic cells and high fences topped with razor wire. When Napuat arrived, the waiters recognized him at once and bustled around deferentially. Once we had settled back in our chairs, I began by asking him about his experiences serving as the minister for internal affairs in the central government but before long he had turned the tables on me. I found myself describing my experiences conducting research with ex-prisoners. I told him that just days previously I had been in a prison in Australia, where senior staff shared with me their personal misgivings about the effectiveness of incarceration as a method of crime prevention or criminal reform. I went on to describe how some of the most tragic stories I had learned from inmates in Queensland related to the treatment of Indigenous Australians.

Napuat nodded in recognition. The European system of criminal justice, he explained, was a bit like a factory production line. When a crime was thought to have occurred, police and lawyers would take months gathering evidence of potential relevance to the case. The process would be hugely expensive and time-consuming. Eventually all the information laboriously gathered would be heard in court, and this too could be a very prolonged process. At long last, a verdict would be reached. If the defendant was found guilty, he or she would usually be sentenced to a term in prison, during which time the prisoner would be fed and housed at great cost to the public purse but without contributing anything to society. And when eventually released, reintegration would typically be difficult. Ex-prisoners would return home to find that their families were suffering, their children neglected, and all these social problems afflicting the innocent were now contributing to further cycles of crime.

The picture Napuat painted sounded both dismal and familiar. But what was the alternative? His answer was to reject the colonial model and adopt a more traditional Melanesian one. According to ancient custom, when somebody committed a crime, everyone would gather without delay in the *nakamal* – the traditional meeting place – to discuss what had happened. Each person would get their chance to speak out – transgressors, victims, and bystanders. The community would listen intently and chip in. After everyone had been heard, the chief would make a ruling. A punishment or compensation payment would be imposed, and the matter would be settled. Even if forgiveness was hard, it had to be worked at. The aim of the traditional system was to restore harmony as rapidly and permanently as possible. Typically, the entire process of establishing guilt and imposing punishment took a single day without requiring a long-drawn-out legal process or a wasteful system of incarceration that only perpetuated the problem. Which system, Napuat wanted to know, did I think was better?

Thanks to Europe's colonial past, virtually every society in the world now operates a version of the long-drawn-out system built around police, lawyers, courts, and prisons. Prisons are so ubiquitous that hardly anyone notices how dysfunctional it is to effectively exclude part of the tribe by locking people up rather than restoring the peace, deterring future transgressions, and bringing ex-offenders back into the fold. And yet if such a system had never existed it is surely inconceivable that we would choose to create it from scratch.

For a start, studies repeatedly suggest that locking people up does not reduce the risk of future criminal behaviour.[18] Statistics aside, however, what studies in criminology seem to show over and over again is that the best way to reduce the risks of reoffending among ex-prisoners is to ensure that they establish positive connections with a law-abiding community – for example by assuming responsibilities in a family, occupation, or voluntary organization. Access to education, healthcare, and housing is obviously also key. As one criminologist put it to me after listing all these factors that are known to reduce recidivism risk, she had yet to read a single study showing that prison was what prevented an ex-offender returning to crime.[19]

Incarceration reduces a person's chances of ever being accepted back into the community – the very opposite of what is needed to prevent future crime. Even before suspects are tried and found guilty, they are often locked up for periods of time. Whether innocent or not, this can have a devastating effect on a prisoner's earnings and relationships, increasing the risk they end up turning to crime in desperation upon their release.[20] Dealing with criminal behaviour – or suspected criminal behaviour – by removing people from the tribe might be less harmful if they at least had supportive families rallying round them. But those who end up in prison commonly lack even this most basic form of social support.

The failure of our current approach to criminal justice is rooted in a failure to appreciate both the power and risks associated with humans' tribal identities. To understand these issues better, I teamed up with a criminologist at the University of Queensland in Australia – Robin Fitzgerald – to measure fusion with law-abiding groups among former prisoners around Brisbane in Queensland. We found that fusion levels with family among probationers were much lower than in the population at large.[21] Also striking were the much lower levels of fusion with country than would be typical of the general population. In interviews, probationers described how their parents beat them, how they would spend long periods living rough on the streets, and how they were estranged from siblings. Given that opportunities among the formerly incarcerated to access education, housing, or jobs were much poorer than for most other people, feelings of alienation are hardly surprising. Participants in our study conveyed their sense of exclusion from the tribe by describing how people on the street would routinely dismiss them as misfits, scarcely disguising their contempt or crossing the road to avoid contact. For many coming out of prison, the nearest thing to family members were other inmates, perhaps the people they had shared a cell with. Such associations are not the ones you need most, however, in order to successfully reintegrate into mainstream society.

These forms of outgroup derogation naturally contribute to recidivism risk among ex-offenders. Former prisoners who end up trapped in a cycle of repeat offending may never find a new tribe to belong to. But even those who do may have further problems. If excluded from mainstream society and even from their own families, the only tribes remaining may be those focused on illegal activities, such as drug-dealing, prostitution, internet fraud, stealing, violence, or terrorism. These tribes don't have careers fairs and open days and they don't advertise for openings on job posting sites or in the newspapers; instead, they are built through relational networks and word of mouth. Prison provides a powerful recruitment ground for such tribes. After all, for those who are isolated and excluded from mainstream society, joining a 'bad' tribe is better than having no tribe at all. In many cases, the tribes available to those with criminal records offer exceptionally potent forms of group bonding, providing a desperately needed alternative to absent or dysfunctional families. When the only groups available are criminal ones, it is hardly surprising that they will easily attract members. There are many street gangs and terrorist organizations that are adept at recruiting outcasts.

For an anthropologist interested in such criminally inclined tribes on the fringes of society, violent football fans provide a particularly attractive focus for research. In part this is because the groups themselves are often legal, even if some of their activities are not. But in part it is because their members are generally easier to access than most terrorist groups, organized criminal gangs, or paramilitary organizations. Moreover, given the wide appeal of football around the world, extreme fan groups can be readily compared across countries. Accordingly, my colleagues and I have been gathering data from extreme football fans across four continents, enabling us to study the role of fusion in motivating illegal behaviours such as fan-based violence. For example, in South America, our research has focused on members of 'superfan' groups known as *torcidas organizadas* – similar to ultras in mainland Europe or hooligan 'firms' in the UK – commonly originating in urban favelas. Although ostensibly focused on supporting their football teams, such groups are also often involved in Brazil's criminal underworld of drugs-trafficking, extortion, and money-laundering, using violence not only as a way of intimidating fans from competing groups but also as a way of maintaining dominance in gang warfare.

Our research has revealed something very important and surprising: criminal violence in superfan groups is not due to maladjustment but misplaced tribalism. Although fans who join violent hooligan groups like these are often portrayed in the media as deviant outsiders, our investigations suggest that the violence associated with extreme fan groups has much more to do with love of the group.[22] Superfans were more violent than ordinary fans by all our measures (which included histories of actual violence as well as expressed willingness to carry out violent acts). Not surprisingly, they were also more highly fused. However, they were not more 'maladjusted'. When we measured their ability to manage interpersonal relationships and work environments or willingness to engage in social and leisure activities with family members, they were no different from ordinary fans. Cleaving more strongly to a tribe – even one that engages in illegal activities – seems to provide better outcomes socially and psychologically than having weaker tribal associations or – worst of all – no tribe at all. What our criminal justice system seems to be doing is to minimize the range of law-abiding groups that ex-prisoners can join and to maximize opportunities to enter violent criminal organizations. It is hard to imagine a more effective way of increasing rates of crime and repeat offending.

All this brings us back to Napuat's question: what kind of criminal justice system do we want? As a member of parliament, Napuat was on a mission to change government policy away from western bureaucratic systems relying on lawyers, judges, and prisons and to move instead in the direction of a traditional approach in which crimes are punished within communities and social harmony rapidly restored. But whatever the merits of Napuat's vision for Vanuatu, I found it hard to imagine it transplanted to my own birthplace, London. That would mean, presumably, creating a *nakamal* for every local neighbourhood and establishing a system of chiefs as well as an elaborated set of rituals and systems of exchange to ensure high levels of community cohesion, cooperation, and consensus – all of which seemed unlikely. But the thought experiment was fruitful in some ways. I found myself reflecting that one feature of traditional Melanesian legal institutions appeared to be patently superior to the ones that had evolved in my own country: the principle that when crimes have been paid for, forgiveness and reinclusion must follow.

Once you are stuck in a western-style criminal justice system, it is hard to get out. Being saddled with a criminal record and released into a world with poor prospects and an inadequate support network, many ex-prisoners are effectively condemned to a life of repeated incarceration. If we want to solve this problem, we need to start by offering them more positive group alignments. Seen in this light, the challenges are twofold. First, how can we encourage ex-prisoners to join law-abiding tribes upon release? And second, how can we improve the willingness of our law-abiding tribes to accept ex-prisoners back into the fold?

One organization in the UK devoted to helping disenfranchised young men find their place in society calls itself ABandOfBrothers. What they offer is a process of initiation into the tribe for ex-prisoners who lack a support network and who typically come from traumatized backgrounds in which addiction and abuse are endemic. The process of initiation itself is fully known only to those who go through it, but the testimony of initiates suggests that the outcome is a powerful bond with each other and with their mentors and a sense of belonging to a larger community that many participants have never experienced before. Those communities are dispersed around the UK but connected through visits and regular communications. Mentors within each of the communities are drawn from a wide variety of backgrounds but what they all have in common is a commitment to help those without a tribe find a home in which they can cooperate and contribute, to support each other, and to learn to stand on their own two feet in a society that had once written them off.

When Conroy Harris, CEO of ABandOfBrothers, visited my office in Oxford, he was immediately drawn to the masks and other artefacts that had been given to me during my fieldwork in Papua New Guinea and he inspected curiously the photos on my walls of events involved in my initiation. Harris explained that he had been born in Antigua but raised in Nottingham, where he joined the RAF at the age of sixteen. After six years' service, he experienced some tough times, including periods of homelessness. But later he developed skills and qualifications in mental health care, discovering in the process that many of the services available were primarily provided for women rather than men. In response to this need, he helped to establish a pioneering group for black men facing mental health challenges. The core ethos of ABandOfBrothers struck me as remarkably similar to the principles of the Kivung – a way of not only initiating young people into the community but also making collective decisions through a process of inclusive discussion and consensus-building in small, interconnected groups. Harris had built a network of highly fused local communities – just like those from which the Kivung movement was composed – in which violence and criminality could be minimized. The key to its success seemed to be that it provided a way of motivating ex-offenders to join a new law-abiding tribe, but it also provided a community through which they could be supported in the long run.

The approach taken by ABandOfBrothers was reminiscent of ancient forms of localized group bonding of a kind that would have been familiar to our Palaeolithic ancestors, as described in Chapter 3. However, I found myself wondering whether an organization of this kind could support much wider forms of extended fusion, rooted in routinized ritual, that could help ex-offenders connect not only with each other and their mentors but with a much larger community.

What would such a community look like? For some, it might be a doctrinal religion of some kind – and many people around the world seeking to reintegrate after release from prison find sanctuary, if only temporarily, in churches, mosques, and temples, and the communities that sustain them. However, many people coming out of prison are not religious and the societies into which they need to integrate may be largely secularized. So, the question we found ourselves asking was whether we could think of another kind of truly large-scale 'imagined community': one associated with values of belonging, loyalty, healthy competition, and family – and which is (mostly) law-abiding?

The answer lay at our feet. People often say that football is their religion, and this is more than just a figure of speech. Football provides the nearest thing many people experience to a sense of loyalty to their country. During periodic regional and global contests, the regalia of national identities are displayed – flags are waved, anthems are sung, heroes are worshipped, and above all common experiences of suffering are created. And all this social glue can be produced by extending fusion generated in much smaller groups – the families and groups of friends who share their love of the beautiful game through experiences of home games in the rain and pilgrimages to away matches in hostile lands. For many, it provides a tribe in much the same way as a religion would, except without the doctrine and ideology. And this makes it better at bringing people from diverse backgrounds and belief systems together, creating the impression of homogeneity by dressing alike, chanting and waving in sync. Perhaps, therefore, football could provide a way of bringing ex-offenders back into the fold of society at large.

It would take a brilliant business leader immersed in the world of football not only to have that thought but to put it into action: David Dein, vice-chairman of Arsenal Football Club from 1983 to 2007 and famous for bringing one of the world's most successful football managers – Arsène Wenger – to the club. Dein's contacts in the world of football enabled him to create a pathbreaking new approach to reintegrating ex-offenders into society and to help them stay out of prison in the long run. He called it the Twinning Project. The aim of the Twinning Project is to twin every prison in

England and Wales (well over a hundred of them) with their local professional football club to bring ex-offenders back into the tribe, to teach them coaching skills that would make them healthier physically and mentally, and to enable them to earn a qualification that could help them find employment after their release from prison. Supported by PE officers in prisons around the UK, professional football coaches would go into prisons twinned with their clubs to help inmates prepare for a better life on the outside after serving their sentences. On 31 October 2018, Dein announced the launch of the Twinning Project to an appreciative audience at Wembley Stadium.

From the moment I first heard about the programme, I was intrigued by how our research on football fandom could help make it as effective as possible. And so I was pleased when, a few weeks after launching the project, Dein came to see me and my colleague Martha Newson in Oxford. She and I had already shown through our research that fusion in football could motivate violence between rival fans, but we also had compelling evidence that it could be used in more positive ways, to build prosocial commitment and help people to live healthier lives, individually and in their relationships with others. The question we started to explore with Dein was whether we could build on this research to help maximize the effectiveness of the Twinning Project. Perhaps we could use the project to harness our in-built tribalism to help ex-prisoners become reintegrated into society at large.

To get started, Martha Newson and I ran a pilot study at five prisons which showed that when inmates increased fusion with the Twinning Project, their case notes improved, and their job attendance increased.[23] We also found evidence that their fusion with criminal groups decreased. The sample size was small (only forty participants), but this was a promising initial foundation on which to build. We set ourselves the task of trying to find out what features of the Twinning Project would best enable participants to reintegrate into society after release from prison. This meant studying the psychological and behavioural effects of participation in the longer run. Fusion is a slow process and 'going straight' means long-term change in thinking and lifestyle.

In particular, we hypothesized that the incorporation of routinized ritual into the Twinning Project would positively impact future-mindedness and help reduce impulsivity among those who had experienced life in prison. We also predicted that the more transformative participants' experiences of the programme were, and the more they attached to the coaches, the more they would fuse to others on the Twinning Project. It will take some years before all the results are in, but our early evidence suggests that the more prison inmates felt they had been personally transformed by the project, the more they became fused to the Twinning Project as a new group identity that mattered to them. Becoming attached to their football coach also increased their fusion with the project.[24] And most importantly, those who participated in the Twinning project were better behaved when they were in prison.[25] Building on these findings, we can generalize what we have learned to other projects as these are rolled out in more countries around the world, and perhaps extended to other kinds of team sports that tap into our tribalistic biases.

Regardless of how successful these kinds of initiatives eventually turn out to be, however, they will not be able to help all or even most people coming out of prisons. If we are to reduce rates of crime, we will need much deeper changes in the way we think about punishment and reform. And that will involve a significant focus not only on ex-prisoners themselves, but on the outlook of the society that is welcoming them back.

And this is the final way that identity fusion might hold the key to a difficult problem. Imagine you were an employer and had to choose between a job candidate with a history of criminal convictions versus somebody with similar qualifications and a clean record. Which person would you hire? It may seem natural to turn down the ex-offender. This is understandable, but for the tribe to flourish we need to work extra hard to hold onto those members most at risk of being lost, if only because the costs of losing them are greater than the costs of keeping them on board. However, if simple arithmetic is not enough, perhaps love for the group might provide the solution. To investigate that possibility, my colleagues and I ran a series of experiments to explore whether shared experiences – one of the key routes to identity fusion – might be enough to persuade prospective employers to offer second chances to ex-offenders.[26] We encouraged would-be employers to believe that fictive candidates for a position with criminal records had undergone personally life-changing experiences that they (the employers) had also endured. The results were consistent and powerful: shared experiences not only increased employers' willingness to hire, but also their willingness to assist with community reintegration efforts and funding.

Such findings hint at what may be the ultimate solution to the savagely high rates of reoffending that plague our society. They indicate that identity fusion – one of the most potent forces of tribalism in human societies – can help motivate ex-offenders to rejoin the tribe. It might also persuade more securely established members of the community to allow them back in.

The Twinning Project likes to quote a phrase from American civil rights activist Jesse Jackson: 'Don't look down on someone unless you are helping them up.' This could be a guiding principle for all the tribes of the world. But it will be a hard message to spread. As it stands, the media helps fuel intertribal conflict by reporting on the salacious details of unusually grisly crimes and stoking the public desire for harsher punishments. Politicians, meanwhile, promise to get tough on crime in part because lurid media reporting feeds the collective appetite for ever-higher rates of incarceration. But most of those locked up around the world are not guilty of the kinds of monstrous acts that make the headlines. Rather than making more arrests and building more prisons it would be in our collective interests to invest in the reform of ex-offenders and in transforming the social conditions that foment criminal behaviour. Excluding and incarcerating members of the tribe is always a sign of failure in a social system.

We have seen in this chapter the Janus face of our inherited forms of tribalism. These are forces that can bring us together but can also lead us into intense forms of outgroup hatred and cruelty – whether by driving social exclusion or group-based violence. And the situation is only getting worse. Throughout history, warfare and militarization helped societies to scale up and compete but that is no longer the case. The largest coalitions created in that way are now too big and powerful to go to war without destroying us all. Does this mean that the time for big tribes has passed? I would argue the opposite. In fact, the time has come for much bigger tribes: ones that are able to bring us together like never before. But this will not be easy. Such a process will require us to tap into all three of our greatest biases at once: conformism, religiosity, and tribalism. And it will require us to do so, for the first time, on a global stage – the focus of my final chapter.

Epilogue: Rise of the Teratribe

People say that the world is getting smaller, but you could also say our groups are getting bigger. Not only because of shared information technologies and population growth, but also because of changes in our mental worlds.

In March 2022, psychology students at the University of Melbourne helped me strap on a virtual reality headset as part of an experiment they had designed to evoke feelings of 'awe' in the lab. As soon as they pressed the start button, I found myself surrounded by other astronauts, gazing through the window of the International Space Station, travelling around 17,500 miles per hour, 250 miles or so above the surface of the earth. At this speed it should have been possible to travel to the moon and back in a single day. Instead, we were circling our home planet. As I gazed upon the scene, the sun suddenly emerged in a burst of light beyond the curved horizon of the earth, like a white diamond. I involuntarily gasped in amazement.

Seeing the world from space, albeit through a virtual reality headset, puts things into perspective. Viewing our planet from afar makes one realize how small it is, how vulnerable to destruction. It becomes easier to appreciate that humankind, for all its seeming diversity, is a single species sharing its home with countless others. And it becomes easier to understand the specialness of humans: that we are the only species on earth capable of understanding what we are looking at from outer space through a VR headset.

The human mind is capable not only of recognizing earth from afar but also of appreciating that the world and its inhabitants are products of history. Even before we understood the basics of plate tectonics or evolution by natural selection, humankind reflected on the question of origins – how we got here and where it is all leading. We can now visualize these processes based on vast amounts of archaeological and historical data at the click of a button on the internet. Anyone who wants to observe the world's civilizations rise and fall on a map of the world – racing through the millennia one year at a time over a period of 200,000 years – can now do so on YouTube in approximately twenty minutes.[1] The technology is undoubtedly extraordinary but so too is the human capacity to recognize our common origins and, above all, to transform this realization into shared *identities*.

Technological innovations have the potential to open our minds to our common humanity in a much more visceral and motivating way than ever before. Not only can we now see more clearly that humankind occupies a single home, but also how global problems are increasingly shaping our collective futures. Today, many of us already grasp this readily when it comes to climate change, biodiversity loss, coral bleaching, and the pollution of our oceans. But our fates are also tethered together when it comes to the way we manage many other issues such as conflict, migration, disease, and poverty. Although our leaders continue to preoccupy themselves with local horizons and short-term goals, the future of our planet depends on being able to think big and act in concert in more globally sustainable ways. But how?

Some of my more specific answers to that question in the previous chapters have tended to focus on relatively parochial problems, such as how to bring about change at the level of communities and nations, and only occasionally at an international level. In this final chapter, however, I want to focus more tightly on what it would mean to draw upon our inheritance more wisely at a global scale. My answer is that we need to work with our natural biases while learning from the lessons of our very unnatural history.

Revolutionizing routines

I've often felt that the founders of the Kivung in East New Britain Province, Papua New Guinea, had a much better understanding of western society than their colonizers did. Unlike the latter, they realized that the activities of missionaries, plantation owners, and colonial authorities amounted to a recipe for environmental disaster. Colonial powers sought to turn the ancient rainforest into a production line for faraway factories, destroying the ancient traditions and lifestyles of indigenous peoples. Followers of the Kivung realized that the newly introduced system of courts and prisons created many social problems, while doing little to deter or reduce crime. And they saw that faith in Christian dogma undermined group identities that were a source of local pride and dignity.

The essence of the Kivung approach, as I explained in earlier chapters, was to unify the rural population behind a new identity and way of life aimed at protecting the rainforest, preventing crime through collective action at the community level, decentralizing decision-making through Melanesian-style citizens' assemblies, and establishing a form of religion that respected tradition while serving the interests of its followers in the here and now. It is an approach that has continued to work for over half a century. While much of Papua New Guinea has faced devastating effects of resource extraction, soaring crime rates, lack of political legitimacy, and gender-based violence resulting from sorcery accusations, followers of the Kivung have been relatively free of these problems.

But perhaps the most impressive aspect of the Kivung is one we have not touched upon before: its openness to new ideas. These collaborative meetings from which the Kivung got its name offered a way of collectively appraising not only the threats that western institutions posed but also the opportunities they might present if adapted to local and regional needs and goals. For all their faults, the models presented by colonial administrations and missions were the product of thousands of years of experimentation, some elements of which had survived as a toolkit for unifying, coordinating, and cooperating on a hitherto unimaginable scale. One of the most powerful new tools in this kit was routinization, and, as we have seen, the Kivung adopted it with relentless enthusiasm: establishing a set of time-consuming daily rituals, announced by the periodic ringing of bells, that helped forge the sprawling Kivung community I encountered in the 1980s. For the world to thrive in the twenty-first century, this openness is the final lesson we must learn from the rainforest. They were open-minded enough to listen respectfully to the Catholic missionaries, but the question is whether we can now be open-minded enough to listen to the wisdom of the Kivung. In order to unite around a set of globally shared beliefs fit for solving universal problems, perhaps we too can embed new routinized behaviours in families, schools, governments, and workplaces – just as members of the Kivung have done.

But if we are to replace our globally shared habits of consumerism with something better, what form should it take? The answer may lie, at least partly, in the necessities of everyday life. When our early Neolithic ancestors first developed methods of crop cultivation and animal domestication, their domestic and community lives became much more routinized. Just as we hoover our carpets and dress ourselves today, early farmers in western Eurasia decorated their dwellings and bodies in distinctive ways. These features of the domestic mode of production – originating in the invention of agriculture – had far greater power over humanity than is generally appreciated. Being houseproud is at least partly about affiliating with the group and avoiding the risk of exclusion by means of wagging tongues: which may be why, for many of us, the hoover gets used more in advance of the arrival of guests than at any other time, and why TV shows devoted to domestic labour from cooking to cleaning are perennially so popular. Domestic rituals are more consequential than we realize; not only could they change climate-impacting behaviour patterns very speedily but perhaps in the process even transform our collective sense of who we are as humans.

Consider our shopping habits. Many of us try to build 'greener' behaviour into our day-to-day lives by shopping more locally, avoiding packaging, reusing bags, and reducing waste. We are all aware that eating less meat, especially beef, reduces our carbon footprints, as does travelling less for family holidays and leisure activities, and especially by flying less frequently. But we don't always follow through. However, all these behaviours can be transformed dramatically when mechanisms of peer pressure come into play. The receipts we are given after making purchases or when we receive fuel bills are an obvious opportunity to provide printed feedback on how our choices are impacting the environment, and how our carbon footprint compares with recognized targets or with other consumers. Although such innovations may be unattractive to retailers, such a system would be no less enforceable than stating the amount of sales tax on each purchase, as is routinely accomplished in many countries.

Norm enforcement often begins at home, not only when values and knowledge are passed down from parent to child but also the other way around. For example, visits to schools by fire fighters are often very effective at improving safety and reducing fire risk at home, because children are excellent vectors for such information in their families. This points to the presence of untapped potential for schools and children's media to spread greener habits in domestic life. Imagine if, instead of children's transnational activism requiring extracurricular or protest-oriented forms of mobilization – as in the case of the School Strike for Climate movement – systems of formal schooling worldwide started to work together in more connected ways to spread ideas and practices that support greener lifestyles, starting in the home. Such initiatives could be fostered in myriad ways by departments of education at national level, building a stronger sense of global community among the young as part of core school curricula worldwide.

Likewise, governments could offer much greater leadership in helping adults directly to reduce their carbon footprints via routinized rituals embedded in domestic life. New habits could be modelled in public institutions, for example by changing patterns of consumption in the heart of government and administration. A first step to removing meat from all cafeterias in the public sector, for example, could be to eliminate it from the menus in the Palace of Westminster or the US Capitol. Or governments could follow the lead of the Scottish parliament, which introduced a raft of measures aimed at reducing their carbon footprint by two-thirds over a twenty-year period, for example by reducing gas and electricity usage and unnecessary business travel.[2]

Historically, transitions on a large scale can be a slow process. In some regions of the world where socio-political complexity emerged independently, the time lag between the initial scaling up of the ritual community and the creation of an enforceable legal framework and system of governance amounted to thousands of years. We cannot afford to take so long. To gain a handle on the challenges facing our planet today, we need to scale up the ritual community from the level of families and nations to that of humanity at large much more rapidly. Fortunately, what we now have – and which our ancient ancestors lacked – is a better scientific understanding of how cooperation works, and the communications infrastructure needed to take advantage of that knowledge. By analysing our own past, for example, we can see that the norms on which larger-scale cooperation was built originated in the routinization of domestic and communal life so we can adapt these to the needs of the world we now find ourselves in.

Routinizing collective habits requires commitment and norm enforcement from the bottom up as well as from the top down. This is not always obvious, however. Whether we live under the thumb of authoritarian regimes or in polarized liberal democracies, we seem to take it for granted that policies should come down to us from on high. But what if it were to be the other way around? What if all policy decisions were made at the bottom of the hierarchy? Decision-making could be the task of thousands of citizens' assemblies – representative samples of the population at large, each saddled with the responsibility to make high-level policy decisions on an assigned cluster of issues in the light of in-depth expert advice, connected into a network of other such decision-making bodies. Such a system could operate in much the same way as jury service – when your number comes up, you are obliged to attend. Ideally, such assemblies would reach their conclusions based on interpersonal interaction and debate, the niche for integrative complexity (IC) and mutual understanding. Such a method of governance could be protected from the negative impacts of news reporting and the polarizing effects of social media. Indeed, just as juries may be insulated from such things, our policymaking assemblies might be similarly sequestered, members surrendering smartphones and laptops at the door as they enter the sanctuary for decision-making, even if deliberations themselves are recorded for wider scrutiny.

In such a system, the role of politicians and governments would look very different – still important but with a new function. Our leaders in parliament would no longer act as the sources of policy but only as elected implementers of it. They would no longer need to produce and defend policies but focus their efforts on demonstrating competence and trustworthiness in the fulfilment of their intended purpose as respected servants of the will of the people, established through well-informed debate. This would still be democracy but routinized along radically new tramlines. Like capitalism, but without its corrosive effects, such a system could spread far and wide, maybe even becoming globalized. Perhaps the assemblies responsible for decision-making could then pool their wisdom to establish a globally accessible fund of collective knowledge on public policy problems and solutions on which we can all draw.

Rethinking world religion

In June 2023, I met Alan Covey – one of my collaborators on the Seshat Global History Databank – at a trendy café in Camden Town to talk about Inca society, a topic on which he is one of the world's greatest experts. Normally to be found surrounded by students in the archaeology department at the University of Texas, he was on one of his annual escapes to London and so it seemed the perfect opportunity to quiz him about the topic of human sacrifice and social inequality in the Inca empire. He wanted me to appreciate that no matter how horrifying the ritualistic killing of children may seem to us, we needed to try to understand it through the lens of Inca cosmology and particularly the idea of reciprocity between gods and mortals.

To make the point even more starkly, Covey pointed out how much more unequal economically our own societies today would have seemed to ancient rulers. He asked me to hazard a guess how many pyramids Elon Musk could afford to build based on today's labour costs. I thought for a moment and concluded that I had no idea. Musk's net worth, Covey continued, was estimated to be around $200 billion; the Great Pyramid of Ancient Egypt was thought to require twenty to forty years for 4,000–5,000 workers to build. Manual workers in Egypt earn around 2,700 Egyptian pounds per month based on 2023 figures and so building the Great Pyramid would cost in labour a maximum of $200 million in today's money. Using his fortune of $200 billion Musk could therefore build 1,000 Great Pyramids.

It is easy to suppose that the pharaohs of Egypt presided over far more unequal societies than those we live in today. In the popular imagination, the gap between rich and poor in ancient times has been closed through the rise of democracy. Nevertheless, according to Oxfam, today's richest 1 per cent own twice as much wealth as the rest of the world put together.[3] It should come as no surprise, therefore, that the yawning chasm between Musk's fortune alone and the savings of most of the people alive today is much greater than the gap between the riches of ancient Egypt's mightiest of pharaohs and lowliest of slaves. And this inequality is not merely economic. The forms of inequality we now endure flagrantly favour the interests of a tiny minority at the expense of society as a whole. For example, Inca rulers demanded, however misguidedly, that humans be sacrificed in order to protect the environment from catastrophic natural disasters. If Elon Musk were obliged to devote even half his fortune to tackling today's climate crisis, imagine how much better off we would be as a global human tribe.

Arguably the reason why it is so hard to imagine the super-rich being required under international law to give up large portions of their wealth to tackle global problems is not because it is impracticable, but because our routinized habits of thought make such ideas almost unthinkable. If the idea of reining in the super-rich sounds like pie in the sky, so too would the idea of a world religion if you tried to explain it to Neolithic farmers. Imagining such things is the first step to making them happen. The Kivung was able to unite formerly warring tribes into a multi-ethnic proto nation in part because its visionary leaders were inspired by the idea of an organized religion and a centralized system of government, modelled by the very colonial system they set out to transcend. Perhaps changing the way we organize ourselves on a global platform would seem less daunting if we were to imagine the existing world religions as a stepping stone in the direction I am proposing.

One way this could happen is to transform established religious organizations into more effective catalysts for global cooperation. All the transnational organized religions, including Christianity, Islam, Judaism, Buddhism, and Hinduism and their myriad offshoots, provide scriptural injunctions to take care of God's creation. But the theological basis for action on the current climate crisis by all devout believers around the world has not been put to much practical use. It could be otherwise. It is not as if these religions are powerless to motivate changes of habit that impact our domestic behaviour. At mealtimes, religious preferences and taboos have long been influencing what we eat. Meanwhile, the homes of the faithful everywhere are adorned with family shrines, sacred iconography, and spaces for sacrifice and prayer that provide opportunities to shape our values and goals – a physical embodiment of the way that faith informs our behaviour in the home. Religion has played this role for millennia. Clean spaces in the Neolithic houses of Çatalhöyük around 8,000 years ago were almost certainly regarded as sacred and richly symbolic – making the house more like a temple in some respects than a mere dwelling.[4] Their clean spaces were rather like the family altars one finds in Catholic homes or the shines in Balinese houses – or indeed the family temples of the Kivung where offerings to the ancestors are laid out. These are all places in which rituals are integrated into everyday life. For those who see themselves as adherents to the world religions – that is, most people on the planet today – life at home is suffused with religious ideas, from child-rearing practices, to daily prayers, and acts of worship.

The fact that these aspects of religious life are not more systematically harnessed in the struggle to tackle environmental issues is a missed opportunity. The world's doctrinal religions have the leadership structures and potential motivational base to implement action plans on climate change and biodiversity loss as well as other major planetary threats. This has been demonstrated by a few high-profile projects, such as the planting of a million trees to mark the 550th anniversary of Sikhism.[5] But such initiatives are much fewer in scope and lower in impact than they could be. Religious organizations routinely impose restrictions on domestic life that have no clearcut scriptural basis – for example on matters of contraception, archaic food taboos, and proper attire – but strikingly neglect obligations of stewardship for the earth, which *does* find extensive support in the texts held most sacred in those traditions. This is a paradox that religious leaders and followers alike could do much more to address. And the more arbitrary religious norms that regulate people's diets could, in theory at least, be augmented quite easily with sufficient will. Why should the norm of eating fish on Fridays in some Catholic communities, for example, not be supplemented by the rule of not eating meat on Mondays?

Just as moralizing religions are failing to utilize our intuitive beliefs with the impact they might, secular environmentalist movements are also failing in comparable ways. For example, my colleagues and I undertook an analysis of the academic literature on wildlife conservation to explore the extent to which intuitive moral arguments were being developed and debated internationally. Following a statistical analysis of the literature, we found that only one of the seven forms of cooperation that humans everywhere consider morally good was being regularly invoked in support of conservation efforts.[6] It is easy to imagine how the environment would benefit if our approach to it were rooted in each of our panhuman moral rules: help your kin, be loyal to your group, reciprocate favours, be courageous, defer to superiors, share things fairly, and respect other people's property. However, the only feature of cooperation widely invoked in the academic literature on conservation turned out to be *reciprocity* – most notably the need to reduce parochial plundering of natural resources with well-enforced reciprocal international agreements. The other six dimensions of morally endorsed cooperation – although obviously relevant to conservation issues – were almost entirely neglected. This too is a missed opportunity.

Religious leaders, parliamentarians, pressure groups, and the mass media could all make a more concerted effort to appeal to our universal instincts. They could appeal to our urge to care for kin by emphasizing the need to safeguard the future for our children and grandchildren. They could appeal to heroism by emphasizing the need to stand up to powerful stakeholders in non-renewables. Principles of fairness are also relevant since we share the earth with millions of other species. And respect for property comes into play since humans only arrived on the scene recently and other species have prior possession of the earth.

Finally, there is much to be gained also from thinking about the way we share information from a more global perspective. I argued in Chapter 8 that transnational corporations, the mass media, and tech platforms are taking over the management of our ancient religious predispositions and susceptibilities in ways that serve vested interests rather than the needs of most individuals or the societies they live in. However, this is not simply a problem for domestic policy to address at a country-wide level. Unfortunately, we have allowed a form of parasitic religiosity to spread worldwide and so it requires a global response. Taking the initiative for behaviour modification away from advertisers and the media and creating a new form of secular religion is a global challenge – one that can only be adequately addressed through regulatory mechanisms built around universal moral principles and worldwide targets for cooperation.

One form this could take is a global news hub funded through the UN or the G20. A global public broadcasting corporation to rival even the most trusted national ones. At the core of it would be an explicitly formulated theory of newsworthiness that prioritizes information about world affairs that we all need to know about in order to be responsible global citizens. Not gossipy, fear-inducing, scandal-mongering, yuck-and-wow entertainment news, interspersed with advertising, but something genuinely useful to understanding the major challenges facing us all, as they unfold day to day.

The global tribe

After darkness fell on 1 July 2015, Cecil was wounded by a single arrow fired by Walter Palmer, an American dentist wielding a longbow. Cecil died of his injuries around twelve hours later. A thirteen-year-old lion living wild in Zimbabwe, Cecil was soon appearing on screens around the world after his death was picked up by the media and went viral. At the time of his death, Cecil was being studied and tracked by some of my friends and colleagues in Oxford. Soon, the university's phone lines were jammed with callers wanting to know what they could do to help protect lions like Cecil in the future.

The research unit at the centre of this storm of public interest was run by David Macdonald. I asked him if we could run a survey with a sample of around a thousand of the much larger number of people donating to lion conservation and he agreed. Over the following six months we gathered longitudinal data on people's reactions to the death of Cecil. What we found was that over time, as people reflected on the incident, they incorporated the tragedy into their personal narratives as a life-changing episode – it became a transformative moment in their views on wildlife conservation. This deep shift in identity was rooted in feelings of shared experience with Cecil himself and, as a result, their feelings of fusion with him and with other wild lions in Africa.[7] This pioneering study demonstrated that identity fusion could occur across the species barrier. If we can fuse with lions, we can fuse with the enormous number of other creatures with whom we share the planet.

The scale of our tribes has expanded greatly over the course of human history, as this book has documented in some detail. But the last bastion in that process of expansion is the tribe of life itself, encompassing not only humanity but all living things. Just as our ancient ancestors initiated us into the responsibilities of adulthood as members of the band, we too could establish forms of initiation for all future generations as custodians of our planet. In the process, we might bequeath not only knowledge of the history of the earth and our place in it, but a growing understanding of how to take care of it and of each other. Ideally, this would take the form of a genuinely transformative rite of passage through which the many privileges of adulthood are achieved rather than simply given. Imagine if eighteen-year-olds across the world had to earn adult status in their societies by performing acts of global citizenship in ways tailored to their country's particular institutional strengths and natural resources. Although performed and completed in local communities, the culminating graduations associated with such contributions to the earth could be celebrated globally in much the same way as we now see in each new year by vicariously sharing its arrival through the time zones on our televisions and computer screens.

The case of Cecil demonstrates quite how far extended fusion can take us. If humans are able to fuse with a lion, then they clearly have the potential to fuse with other humans – however different they might seem. And this raises an interesting question: could humanity at large begin to see itself as a single tribe? When I suggest to my students that the most neglected tribe in the world is humanity itself, they usually smile indulgently before arguing that such a form of unification would require a common enemy. Inevitably, the subject at some point turns to Martian invasions and other sci-fi fantasies. However, this is not what real science teaches us. As

I explained in Chapter 6, *identification* with large group categories may indeed entail a competitive attitude towards outgroups. But this is not a necessary feature of fusion. When we fuse with a group, it is more like the feelings of love we have for our families, or maybe even for strangers at an overnight rave. We are strengthened by the group, and we give it strength in return. This is an indefinitely extendable form of loyalty. There is no need for an outgroup or enemy. Only when the beloved group comes under attack do we discover a dark side to fusion – the willingness to fight and die if necessary to bring down the hated oppressor. As long as there is no external threat, however, the default position of fused groups is to cooperate with others peacefully.

To investigate the feasibility of fusion with humanity at large, I teamed up with Lukas Reinhardt, a doctoral student in the department of economics at the University of Cologne with an interest in the concept. We agreed that the hypothesis was reasonable, but more evidence was needed to test it. We decided to focus on the two pathways to fusion discussed at length in earlier chapters: shared biology and shared self-defining experiences. To explore the effects of shared biology we elected to show a sample of ordinary people from the general population a YouTube video in which popular author A. J. Jacobs invited everybody in the world to a family reunion.[8] In the clip, Jacobs argued that we all trace descent from a common human ancestor and are therefore more closely related biologically than we may have realized. Reinhardt and I wanted to find out how people who watched this video scored afterwards on a measure of fusion with humanity, and in particular whether it made them more generous to people from other countries. In the case of the shared experiences pathway, we focused on a large sample of mothers. Would those whose lives had been most transformed by the experience of childbirth be more fused with other mothers around the world? In both cases, the answer was a resounding yes. Simply being reminded of those aspects of our personal essence that can be shared with the rest of humanity makes us more fused and therefore more willing to help solve global problems.[9]

What is true of the general population should also be true of those who wield power: heads of corporations, politicians, religious leaders, and other shapers of public opinion. When one thinks of the leaders who command the greatest respect on a global platform, they are not necessarily the ones who lead the richest or most powerful states. Nelson Mandela was a remarkable example of this – an individual whose influence derived not from economic or geopolitical leverage but from the ability to recognize shared experiences of humanity, even to appreciate the power of such shared experiences with those who harmed him personally and oppressed the communities of South Africa in which he was raised. Such leaders may be described as 'barrier crossers'.[10]

Barrier crossers are those who recognize that simply pursuing the interests of the ingroup in competition with outgroups often carries heavy costs for everyone. By working together, however, we can all be better off. Of course, working with one's enemies is difficult. Lack of trust is a common impediment. So too, unfortunately, is the prospect of seeing outgroups thrive. Indeed, for many passionate supporters of an ingroup, the desire to harm an outgroup can be even stronger than the desire to benefit the ingroup. This is obviously a worrying problem, but barrier-crossing leadership can provide a solution.

In one study,[11] we recruited sixty leaders engaged in three kinds of embattled communities: African-American leaders from US cities with a long history of racial divisions, Irish Traveller leaders in Dublin, and leaders of minority Muslim communities in London. Thirty-three leaders in our sample were identified in advance of the study as barrier crossers, due to their longstanding deep engagement with members of outgroups as well as fellow ingroup members to help resolve problems of mutual concern, and twenty-seven were identified as barrier-bound leaders who worked mainly or exclusively with ingroup members to advance their collective interests. The goal of our study was to explore some of the underlying psychological orientations that might help to explain these two quite different approaches to leadership. It seemed plausible, for example, that empathizing ability might account for the ability of barrier-crossing leaders to work with outgroups. We found no evidence for that, however. What we did find was that barrier crossers recognized that our most transformative experiences, especially experiences of suffering, are typically shared across groups as well as within them. This was an important finding because it indicated that fusion could be extended beyond the ingroup more readily among the barrier crossers – by appreciating that certain kinds of experiences and memories transcend group boundaries.

This points to one of the most fundamental problems that has continually resurfaced throughout this book: how to cultivate forms of leadership capable of turning the herd in new directions and appealing to popular instincts in ways that are beneficial to society at large and not just to the interests of manipulative elites. Barrier-crossing leadership offers a potential answer to this puzzle. What makes identity fusion such an asset to leaders of social change is not only that it motivates pro-group action but that it does so fearlessly in the face of resistance and unpopularity: it motivates people to do what they believe is best for the group whatever the personal cost. Barrier-crossing leaders do the same. But they do so with a more expanded concept of the group in mind.

We could describe this type of leadership in another word: courageous. If we are to tackle the problems of environmental degradation, societal breakdown, and violent conflict that bedevil the world we live in today, we need leaders who are not only intelligent and strategic but also brave.

Seeing the earth from afar through a virtual reality headset made me realize in a more visceral way than ever before that humanity is a single tribe, dependent on a single source of finite resources. And yet viewing our home from outer space is not enough. We also need to develop a much deeper understanding of the constraints of human nature and the lessons of our history if we are to build the future we seek.

Ten thousand years ago, the human tribe was limited to the people one knew personally. By the time the first states appeared, much of humanity had become divided into 'kilotribes' composed of thousands of faceless others who shared the same basic beliefs and practices but still excluded and dehumanized peoples regarded as falling outside that category. The Axial Age marked a transition to megasocieties, by crossing the threshold of around a million individuals. Humanity is arguably now on the cusp of becoming a 'teratribe', in which the billions now living on the earth finally unify to address issues affecting us all. We may think it is more natural to see ourselves as the citizens of nation states, but this is due to habits of thinking rather than some more insuperable impediment.

The price of failing to scale up is clear. Even though we produce enough food to feed everyone on the planet, millions go hungry; even though we know that all-out nuclear war would destroy us all, we continue to stockpile deadly weaponry; even though we know that the planet is heating up, sea levels rising, and finite resources are running out, we continue to emit greenhouse gases at an unsustainable rate. Just as the great T-shaped monoliths at Göbekli Tepe may have been the last hurrah of a Palaeolithic way of life that was on the way out, the excessive consumerism and populism that defines our age may one day be interpreted as the death throes of a phase in human history when we finally surmounted the threshold of a new form of global unification. Humanity will no doubt always be subdivided into innumerable cultural groups defined by distinct cultures, languages, dialects, and norms. But perhaps we are at last entering a stage in our evolutionary history when we can also pull together effectively at a higher level when we need to.

This book has focused on three dimensions of human nature seen from three distinct perspectives: the biological evolution of our psychology over millions of years, the cultural evolution of our political and economic systems over thousands of years, and the problems we are now facing as a result. My central argument has been that although our ability to harness and manage the three biases allowed us to scale up cooperation in the past, our inherited methods of cooperation are now herding us down a pathway to destruction. But by understanding better how cultural evolution allowed us to overcome the limitations of human nature in the past, we can use that knowledge to make game-changing decisions about our collective futures. Only some of the old methods remain viable. Others do not, and it is crucial to recognize which is which. We need to take a series of steps that will motivate us to shape the world we live in and plan for its future in ways that are both deliberate and consensual rather than haphazard and divisive.

Above all, it is crucial that every community on earth is prepared to learn from every other. As well as drawing on the rich resources of academic research, we can learn from the insights of indigenous groups into the ills of global capitalism and how we might address them. The idea that theories of 'development' should invariably flow from the affluent west to the poor and uneducated is not just arrogant and condescending, it sells us all short. What we need is a global debate rather than a lecture from the west to the rest. My mentors in the Kivung were fascinated by questions about human origins. Since most people in my village lacked literacy skills and access to books, they developed their own tentative theories. But these were always presented to me with a gentle curiosity, rather than a dogmatic insistence on any one interpretation. This is exactly the spirit in which cooperation can be scaled up.

The earth is four-and-a-half billion years old, and – barring human-induced catastrophe – may be potentially habitable for another billion years or so. With such vast time depths in mind, the appearance of our species is no more than the blink of an eye. And yet, as technological advancements outpace our social instincts, the future of our planet now hangs in the balance. Can our capacities for large-scale cohesion and cooperation catch up with our destructive impulses to plunder and destroy? Can the psychology of our foraging ancestors be adapted to a rapidly changing world in which conformism, religiosity, and tribalism can work for us instead of against us? If we draw on the wisdom of the past and the science of the present – the fruits of our unnatural history – we can safeguard the future both of humanity and of a world that now depends, as never before, on how wisely we draw upon our collective inheritance.

Acknowledgements

This book is the result of nearly four decades of research and collaboration, during which incalculable debts have inevitably accumulated. Supervising over thirty PhD students and over forty postdocs and collaborating with countless colleagues at universities all around the world has forced me to learn far more than I imagined myself capable, for which I am humbly grateful. Since this book trespasses on the expertise of so many specialists, I have a long list of colleagues to thank warmly for reading and commenting on excerpts bearing on their field of expertise (while acknowledging of course that responsibility for any remaining errors lies with me): Scott Atran, Pascal Boyer, Michael Buhrmester, Emma Cohen, Alan Covey, Julia Ebner, Kevin Foster, Pieter François, Peter Frankopan, Stewart Guthrie, Conroy Harris, Ian Hodder, Dan Hoyer, Robert Jagiello, Chris Kavanagh, Jack Klein, Ian Kuijt, Jennifer Larson, Robert N. McCauley, Michal Misiak, Martha Newson, Kate Raworth, Lukas Reinhardt, Ralph Schroeder, Paul Seabright, Alan Strathern, Peter Turchin, Valerie van Mulukom, Claire White, and Fiona White. In addition, I should like to thank my research assistant, Danielle Morales, for fact-checking many pages of cited material.

Portions of the text in Chapters 1, 3, and 4 were first published in *Aeon* magazine in Whitehouse, H. (2012), 'Human Rites' (available from: https://aeon.co/essays/rituals-define-us-in-fathoming-them-we-might-shape-ourselves) and portions of material in Chapter 3 were first published *in Pacific Standard* magazine in Whitehouse, H. (2016), 'What Motivates Extreme Self-sacrifice' (available from: https://psmag.com/social-justice/what-motivates-extreme-self-sacrifice).

I am also indebted to Rowan Borchers, my editor at Penguin Random House, for encouraging me to write the book in the first place and for pulling apart the first draft I submitted over a year later and helping me see how best to reassemble it, incorporating also the views of Rachel Field at Harvard University Press. Despite having to be persuaded to make some painful cuts to a manuscript that grew out of control, I learned a lot from the process and developed a profound admiration for Rowan's and Rachel's professional insights and skills. I would also like to thank the wider editorial team at PRH – including the copy-editor Lindsay Davies – for all their sharp observations.

Above all, I would like to thank my wife, Merridee, for reading and commenting on entire drafts of the book, for providing boundless emotional support when my morale dipped, and for patiently tolerating my spells of self-absorption when it seemed that my 'ear flaps were down' or I was 'thinking too deeply'. I am grateful also to my son Danny, my daughter-in-law Sally, and my mother Patricia for their many curious questions throughout the writing process. I have dedicated this book to all of them, together with my granddaughter Delilah, whose future depends on the speed with which we can rescue our ailing civilization. The support of my family throughout the writing process – including my wonderful extended family in Australia – has meant more than I can say. One of the core themes of this book is that just as bonds of kinship have been central to the evolution of human civilization, they may be one of our most powerful means of saving it. Although this is a work that has no regard for academic silos, it is written with the deepest respect and love for my fellow human beings. We all descend from common ancestors and so we are all, in a very real sense, one family.

Notes

Introduction

1 McCauley, Robert N., *Why Religion Is Natural and Science Is Not*, Oxford University Press (2013).
2 Pinker, Steven, *The Blank Slate: The Modern Denial of Human Nature*, Penguin (2003).
3 Tooby, J. & Cosmides, L., 'The Psychological Foundations of Culture', in Barkow, J., Cosmides, L., & Tooby, J. (Eds.), *The Adapted Mind: Evolutionary Psychology and the Generation of Culture*, Oxford University Press (1992).
4 Singer, Peter, *A Darwinian Left: Politics, Evolution, and Cooperation*, Yale University Press (2000).
5 They are also entwined at multiple levels – including the expression of genes, the maturation of beliefs, and the evolution of entire social systems (see Chapter 5 of my book *The Ritual Animal: Imitation and Cohesion in the Evolution of Social Complexity*, Oxford University Press, 2021).
6 Povinelli, Daniel, *Folk Physics for Apes: The Chimpanzee's Theory of How the World Works,* Oxford University Press (2003).
7 Meng, X., Nakawake, Y., Hashiya, K., Burdett, E., Jong, J., & Whitehouse, H., 'Preverbal Infants Expect Agents Exhibiting Counterintuitive Capacities to Gain Access to Contested Resources', *Scientific Reports*, Vol. 11, No. 10884 (May 2021). DOI: 10.1038/s41598-021-89821-0
8 McKay, R., Herold, J., & Whitehouse, H., 'Catholic Guilt? Recall of Confession Promotes Prosocial Behavior', *Religion, Brain & Behavior*, Vol. 3, No. 3, pp. 201–9 (2013).
9 Atkinson, Q. D. & Whitehouse, H., 'The Cultural Morphospace of Ritual Form: Examining Modes of Religiosity Cross-culturally', *Evolution and Human Behavior*, Vol. 32, No. 1, pp. 50–62 (2011).

10 For example, see Kapitány, R., Kavanagh, C., & Whitehouse, H., 'Ritual Morphospace Revisited: The Form, Function and Factor Structure of Ritual Practice', *Philosophical Transactions of the Royal Society B*, Vol. 375, No. 1805 (2020). DOI: 10.1098/rstb.2019.0436
11 Whitehouse, Harvey, *The Ritual Animal*.

1. *Copycat Culture*

1 Legare, C. H., Wen, N. J., Herrmann, P. A., & Whitehouse, H., 'Imitative Flexibility and the Development of Cultural Learning', *Cognition*, Vol. 142, pp. 351–61 (2015).
2 Whitehouse, Harvey, *The Ritual Animal: Imitation and Cohesion in the Evolution of Social Complexity*, Oxford University Press (2021).
3 The BBC website provides links to clips from the series here, on which the following descriptions are based: www.bbc.co.uk/programmes/p06d1n2f/clips
4 'Bhumi's Ultimate Test of Faith', *Extraordinary Rituals*, Series 1, Episode 3, 'Changing World', BBC2 (2018): www.bbc.co.uk/programmes/p06j1h6q
5 Whitehouse, H., 'The Coexistence Problem in Psychology, Anthropology, and Evolutionary Theory', *Human Development*, Vol. 54, pp. 191–9 (2011); Jagiello, R., Heyes, C., & Whitehouse, H., 'Tradition and Invention: The Bifocal Stance Theory of Cultural Evolution', *Behavioral and Brain Sciences*, Vol. 45 (2022). DOI: 10.1017/S0140525X22000383
6 Gergely, G. & Csibra, G., 'Sylvia's Recipe: The Role of Imitation and Pedagogy in the Transmission of Cultural Knowledge', in Enfield, N. J. & Levenson, S. C. (Eds.), *Roots of Human Sociality: Culture, Cognition, and Human Interaction*, Berg Publishers (2006).
7 This, at least, was common practice when I was training to become an anthropologist in the 1980s. Over the decades since, it has become increasingly common for anthropologists to study the communities they themselves were raised in, or at least have some lifelong connections with. Although this can have many benefits (e.g., linguistic fluency and strong connections in the community being studied), it can also make it harder to view what people say and do with the fresh eyes of a newcomer.

8 Wilson, D. S. & Whitehouse, H., 'Developing the Field Site Concept for the Study of Cultural Evolution (with Comment)', *Cliodynamics*, Vol. 7, No. 2, pp. 228–87 (2016).

9 Meltzoff, A. N., Waismeyer, A., & Gopnik, A., 'Learning About Causes from People: Observational Causal Learning in 24-Month-Old Infants', *Developmental Psychology*, Vol. 48, No. 5, pp. 1215–28 (2012).

10 Von Bayern, A. M. P., Heathcote, R. J. P., Rutz, C., & Kacelnik, A., 'The Role of Experience in Problem Solving and Innovative Tool Use in Crows', *Current Biology*, Vol. 19, No. 22, pp. 1965–8 (2009).

11 This is not to say that chimps *never* overimitate – see Whiten, A., McGuigan, N., Marshall-Pescini, S., & Hopper, L. M., 'Emulation, Imitation, Over-imitation and the Scope of Culture for Child and Chimpanzee', *Philosophical Transactions of the Royal Society B*, Vol. 364, No. 1528, pp. 2417–28 (2009).

12 Nagell, K., Olguin, R. S., & Tomasello, M., 'Processes of Social Learning in the Tool Use of Chimpanzees (*Pan troglodytes*) and Human Children (*Homo sapiens*)', *Journal of Comparative Psychology*, Vol. 107, No. 2, pp. 174–86 (1993).

13 Lyons, D. E., Damrosch, D. H., Lin, J. K., Macris, D. M., & Keil, F. C., 'The Scope and Limits of Overimitation in the Transmission of Artifact Culture', *Philosophical Transactions of the Royal Society B, Biological Sciences*, Vol. 366, No. 1567, pp. 1158–67 (2011).

14 Lyons, D. E., Young, A. G., & Keil, F. C., 'The Hidden Structure of Overimitation', *Proceedings of the National Academy of Sciences USA (PNAS)*, Vol. 104, No. 50, pp. 19751–6 (2007).

15 Henrich, J. & Gil-White, F. J., 'The Evolution of Prestige: Freely Conferred Deference as a Mechanism for Enhancing the Benefits of Cultural Transmission', *Evolution and Human Behavior*, Vol. 22, No. 3, pp. 165–96 (2001).

16 Lyons, D. E., et al., 'The Hidden Structure of Overimitation'.

17 Gergely, G., Bekkering, H., & Király, I., 'Rational Imitation in Preverbal Infants', *Nature*, Vol. 415, No. 755 (2002). DOI: 10.1038/415755a

18 Hoffer, Eric, *The Passionate State of Mind and Other Aphorisms*, Harper & Bros. (1955).

19 Herrmann, P. A., Legare, C. H., Harris, P. L., & Whitehouse, H., 'Stick to the Script: The Effect of Witnessing Multiple Actors on Children's Imitation', *Cognition,* Vol. 129, No. 3, pp. 536–43 (2013).

20 Watson-Jones, R. E., Legare, C. H., Whitehouse, H., & Clegg, J., 'Task-Specific Effects of Ostracism on Imitation of Social Convention in Early Childhood', *Evolution and Human Behaviour*, Vol. 35, No. 3, pp. 204–10 (2014).

21 Watson-Jones, R. E., Whitehouse, H., & Legare, C. H., 'In-group Ostracism Increases High Fidelity Imitation in Early Childhood', *Psychological Science*, Vol. 27, No. 1 (2015). DOI: 10.1177/0956797615607205

22 Williams, K. D. & Jarvis, B., 'Cyberball: A Program for Use in Research on Interpersonal Ostracism and Acceptance', *Behavior Research Methods*, Vol. 38, No. 1, pp. 174–80 (2006).

23 Milgram, Stanley, 'Behavioral Study of Obedience', *Journal of Abnormal and Social Psychology*, Vol. 67, No. 4, pp. 371–8 (1963).

2. Wild Religion

1 Whitehouse, Harvey, *Inside the Cult: Religious Innovation and Transmission in Papua New Guinea*, Oxford University Press (1995).

2 The Baining comprised several language groups – including the Mali – but since the people in my village typically presented themselves to outsiders using the more inclusive term 'Baining', I will stick with this catch-all term throughout the book.

3 The full title of the movement is Pomio Kivung, in acknowledgement of the fact that it encompassed numerous language groups in the Pomio region where its headquarters were established and which is some considerable distance from the lands of the Baining.

4 Whitehouse, Harvey, *Arguments and Icons: Divergent Modes of Religiosity*, Oxford University Press (2000).

5 Whitehouse, H., 'Apparitions, Orations, and Rings: Experience of Spirits in Dadul', in Mageo, J. M. & Howard, A. (Eds.), *Spirits in Culture, History, and Mind*, Routledge (1996).

6 Cohen, Emma, *The Mind Possessed: The Cognition of Spirit Possession in an Afro-Brazilian Religious Tradition*, Oxford University Press (2007).
7 Boyer, P., 'Informal Religious Activity Outside Hegemonic Religions: Wild Traditions and Their Relevance to Evolutionary Models', *Religion, Brain & Behavior*, Vol. 10, No. 4, pp. 459–72 (2019).
8 Whitehouse, Harvey, *Modes of Religiosity: A Cognitive Theory of Religious Transmission*, AltaMira Press (2004).
9 Heyes, Cecilia, *Cognitive Gadgets: The Cultural Evolution of Thinking*, Harvard University Press (2018).
10 In addition to Pascal Boyer and me, this group included: E. Thomas Lawson, Robert N. McCauley, Justin Barrett, and Brian Malley. See Whitehouse, H., 'Twenty-Five Years of CSR: A Personal Retrospective', in Martin, Luther H. & Wiebe, Donald (Eds.), *Religion Explained? The Cognitive Science of Religion After Twenty-Five Years*, Bloomsbury Academic (2017).
11 White, C., Barrett, J. L., Boyer, P., Lawson, E. T., McCauley, R. N., & Whitehouse, H., 'The Cognitive Science of Religion: Past, Present, and Possible Futures', *Religion, Brain & Behavior* (in press).
12 Baron-Cohen, S., 'How to Build a Baby That Can Read Minds: Cognitive Mechanisms in Mindreading', *Cahiers de Psychologie Cognitive/Current Psychology of Cognition*, Vol. 13, No. 5, pp. 513–52 (1994).
13 One of the earliest methods of testing theory of mind revealed that autistic children had difficulties in mind-reading tasks that most children found easy. See Baron-Cohen, S., Leslie, A. M., & Frith, U., 'Does the Autistic Child Have a "Theory of Mind"?', *Cognition*, Vol. 21, No. 1, pp. 37–46 (1985).
14 Bering, J. M., 'The Folk Psychology of Souls', *Behavioral and Brain Sciences*, Vol. 29, No. 5, pp. 453–98 (2006). http://cognitionandculture.net/wp-content/uploads/10.1.1.386.3734.pdf
15 Bering, J. M. & Bjorklund, D. F., 'The Natural Emergence of Reasoning About the Afterlife as a Developmental Regularity', *Developmental Psychology*, Vol. 40, No. 2, pp. 217–33 (2004).
16 Boyer, P. & Liénard, P., 'Why Ritualized Behavior? Precaution Systems and Action Parsing in Developmental, Pathological and Cultural Rituals', *Behavioral and Brain Sciences*, Vol. 29, No. 6, pp. 595–613 (2006).

17 Fiske A. P. & Haslam, N., 'Is Obsessive-Compulsive Disorder a Pathology of the Human Disposition to Perform Socially Meaningful Rituals? Evidence of Similar Content', *Journal of Nervous and Mental Disease*, Vol. 185, No. 4, pp. 211–22 (1997).

18 Nemeroff, C. & Rozin, P., 'The Contagion Concept in Adult Thinking in the United States: Transmission of Germs and of Interpersonal Influence', *Ethos*, Vol. 2, No. 2, pp. 158–86 (1994).

19 Frazer, Sir James George, *The Golden Bough: A Study in Magic and Religion*, Oxford University Press (1890, this ed. 2009); Douglas, Mary, *Purity and Danger: An Analysis of Concepts of Pollution and Taboo*, Routledge (2002); Hood, Bruce, *Supersense: From Superstition to Religion – The Brain Science of Belief*, Constable (2009); Bastian, B., Bain, P., Buhrmester, M. D., et al., 'Moral Vitalism: Seeing Good and Evil as Real, Agentic Forces', *Personality and Social Psychology Bulletin*, Vol. 41, No. 8, pp. 1069–81 (2015).

20 Hood, Bruce, *Supersense: From Superstition to Religion*.

21 Bastian, B., Vauclair, C-M., Loughnan, S., et al., 'Explaining Illness with Evil: Pathogen Prevalence Fosters Moral Vitalism', *Proceedings of the Royal Society B*, Vol. 286, No. 1914 (2019). DOI: 10.1098/rspb.2019.1576

22 Kundert, C. & Edman, L. R. O., 'Promiscuous Teleology: From Childhood Through Adulthood and From West to East', in Hornbeck, R., Barrett, J., & Kang, M. (Eds.), *Religious Cognition in China: New Approaches to the Scientific Study of Religion*, Vol. 2, Springer (2017).

23 Kelemen, D., 'Why Are Rocks Pointy? Children's Preference for Teleological Explanations of the Natural World', *Developmental Psychology*, Vol. 35, No. 6, pp. 1440–52 (1999).

24 Barrett, Justin L., *Why Would Anyone Believe in God?*, AltaMira Press (2004).

25 White, Claire, *An Introduction to the Cognitive Science of Religion: Connecting Evolution, Brain, Cognition and Culture*, Routledge (2021).

26 Boyer, Pascal, *Religion Explained: The Evolutionary Origins of Religious Thought*, Basic Books (2001).

27 Hespos, Susan, 'Physics for Infants: Characterizing the Origins of Knowledge About Objects, Substances, and Number', *Wiley Interdisciplinary Reviews: Cognitive Science*, Vol. 3, No. 1, pp. 19–27 (2012).

28 Stahl, A. E. & Feigenson, L., 'Observing the Unexpected Enhances Infants' Learning and Exploration', *Science*, Vol. 348, No. 6230, pp. 91–4 (2015); Köster, M., Langeloh, M., & Hoehl, S., 'Visually Entrained Theta Oscillations Increase for Unexpected Events in the Infant Brain', *Psychological Science*, Vol. 30, No. 11, pp. 1656–63 (2019).

29 Barrett, J. & Nyhof, M., 'Spreading Non-natural Concepts', *Journal of Cognition and Culture*, Vol. 1, No. 1, pp. 69–100 (2001); Boyer, P. & Ramble, C., 'Cognitive Templates for Religious Concepts: Cross-cultural Evidence for Recall of Counter-intuitive Representations', *Cognitive Science*, Vol. 25, No. 4, pp. 535–64 (2001).

30 Boyer, Pascal, *Religion Explained*.

31 Boyer, P. & Ramble, C., 'Cognitive Templates for Religious Concepts'.

32 Barrett, J. L., 'The (Modest) Utility of MCI Theory', *Religion, Brain & Behavior*, Vol. 6, No. 3, pp. 249–51 (2016); for a critical overview of the first couple of decades of research on MCI theory, see Purzycki, B. G. & Willard, A. K., 'MCI Theory: A Critical Discussion', *Religion, Brain & Behavior*, Vol. 6, No. 3, pp. 207–48 (2015).

33 Kapitany, R., 'Why Children Really Believe in Santa: The Surprising Psychology Behind Tradition', *The Conversation* (2019). https://theconversation.com/why-children-really-believe-in-santa-the-surprising-psychology-behind-tradition-126783

34 Meng, X., Nakawake, Y., Hashiya, K., Burdett, E., Jong, J., & Whitehouse, H., 'Preverbal Infants Expect Agents Exhibiting Counterintuitive Capacities to Gain Access to Contested Resources', *Scientific Reports*, Vol. 11, No. 10884 (2021). DOI: 10.1038/s41598-021-89821-0

35 Pew Research Center, 'Pew Research Global Attitudes Project' (2007). Available at: www.pewglobal.org/2007/10/04/chapter-3-views-of-religion-and-morality/

36 McKay, R. & Whitehouse, H., 'Religion and Morality', *Psychological Bulletin*, Vol. 141, No. 2, pp. 447–73 (2015).

37 *Plato: Euthyphro, Apology, Crito, Phaedo*, Greek with translation by Chris Emlyn-Jones and William Preddy, Loeb Classical Library 36, Harvard University Press (2017).

38 Curry, O. S., 'Morality as Cooperation: A Problem-Centred Approach', in Shackelford, T. K. & Hansen, R. D. (Eds.), *The Evolution of Morality*, Springer International Publishing (2016); Curry, O. S., Jones Chesters, M., & Van Lissa, C. J., 'Mapping Morality with a Compass: Testing the Theory of "Morality-as-Cooperation" with a New Questionnaire', *Journal of Research in Personality*, Vol. 78, pp. 106–24 (2019).

39 De Waal, Frans, *Good Natured: The Origins of Right and Wrong in Humans and Other Animals*, Harvard University Press (1996); Dugatkin, Lee Alan, *Cooperation Among Animals: An Evolutionary Perspective*, Oxford University Press (1997).

40 Curry, O. S. (2016). 'Morality as Cooperation: A Problem-Centred Approach'. In Shackelford, T. K. & Hansen, R. D. (Eds.), *The Evolution of Morality*, Springer International Publishing (2016), pp. 27–51. DOI: 10.1007.

41 Curry, O. S., Mullins, D. A., & Whitehouse, H., 'Is It Good to Cooperate? Testing the Theory of Morality-as-Cooperation in 60 Societies', *Current Anthropology*, Vol. 60, No.1 (2019). DOI: 10.1086/701478

42 Lagacé, R. O., 'The HRAF Probability Sample: Retrospect and Prospect', *Cross-cultural Research*, Vol. 14, No. 3, pp. 211–29 (1979).

43 Evans-Pritchard, E. E., *Witchcraft and Oracles Among the Azande*, Oxford University Press (1937, this ed. 2002).

44 Boyer, P., 'Informal Religious Activity Outside Hegemonic Religions: Wild Traditions and Their Relevance to Evolutionary Models', *Religion, Brain & Behavior*, Vol. 10, No. 4, pp. 459–72 (2019).

3. Social Glue

1 McQuinn, B., 'History's Warriors: The Emergence of Revolutionary Brigades in Misrata', in Cole, Peter & McQuinn, Brian (Eds.), *The Libyan Revolution and Its Aftermath*, Oxford University Press (2015).

2 Maynard Smith, J., 'Group Selection and Kin Selection', *Nature*, Vol. 201, No. 4924, pp. 1145–7 (1964).

3 Rowley, I., '"Rodent-Run" Distraction Display by a Passerine, the Superb Blue Wren *Malurus Cyaneus* (L.)', *Behaviour*, Vol. 19, No. 1–2, pp. 170–6 (1962).

4 Whitehouse, H., Jong, J., Buhrmester, M. D., Gómez, Á., Bastian, B., Kavanagh, C. M., Newson, M., Matthews, M., Lanman, J. A., McKay, R., & Gavrilets, S., 'The Evolution of Extreme Cooperation via Shared Dysphoric Experiences', *Scientific Reports*, Vol. 7, No. 44292 (2017). DOI: 10.1038/srep44292
5 Whitehouse, H., Kahn, K., Hochberg, M. E., & Bryson, J. J., 'The Role for Simulations in Theory Construction for the Social Sciences: Case Studies Concerning Divergent Modes of Religiosity', *Religion, Brain & Behavior*, Vol. 2, No. 3, pp. 182–201 (2012).
6 *Extraordinary Rituals*, Series 1, 'Changing World', BBC2 (2018). Available at: www.bbc.co.uk/programmes/articles/1JmvdLwyr5vYH7nzjmmBhL7/why-would-you-do-this
7 Lewis, Gilbert, *Day of Shining Red: An Essay on Understanding Ritual*, Cambridge University Press (1980, this ed. 2008).
8 Whitehouse, H., 'Rites of Terror: Emotion, Metaphor and Memory in Melanesian Initiation Cults', *Journal of the Royal Anthropological Institute*, Vol. 2, No. 4, pp. 703–15 (1996); Xygalatas, Dimitris, *Ritual: How Seemingly Senseless Acts Make Life Worth Living*, Profile Books (2022).
9 Sosis, R., Kress, H. C., & Boster, J. S., 'Scars for War: Evaluating Alternative Signaling Explanations for Cross-cultural Variance in Ritual Costs', *Evolution and Human Behavior*, Vol. 28, No. 4, pp. 234–47 (2007).
10 Buhrmester, M., Zeitlyn, D., & Whitehouse, H., 'Ritual, Fusion, and Conflict: The Roots of Agro-pastoral Violence in Rural Cameroon', *Group Processes & Intergroup Relations*, Vol. 25, No. 1 (2020). DOI: 10.1177/1368430220959705
11 For further details, see Whitehouse, H. & McQuinn, B., 'Divergent Modes of Religiosity and Armed Struggle', in Juergensmeyer, M., Kitts, M., & Jerryson, M. (Eds.), *The Oxford Handbook of Religion and Violence*, Oxford University Press (2012).
12 Maclure, R., '"I Didn't Want to Die So I Joined Them": Structuration and the Process of Becoming Boy Soldiers in Sierra Leone', *Terrorism and Political Violence*, Vol. 18, No. 1, pp. 119–35 (2006).
13 'Road to Heaven, Taiwan', *Extraordinary Rituals*, Series 1, 'Changing World', BBC2 (2018). Available at: www.bbc.co.uk/programmes/p06j1h71
14 Swann, W. B., Jr. & Buhrmester, M., 'Identity Fusion', *Current Directions in Psychological Science*, Vol. 24, No. 1, pp. 52–7 (2015).

15 Swann, W. B., Jr., Gómez, Á., Dovidio, J., Hart, S., & Jetten, J., 'Dying and Killing for One's Group: Identity Fusion Moderates Responses to Intergroup Versions of the Trolley Problem', *Psychological Science*, Vol. 21, No. 8, pp. 1176–83 (2010).

16 Swann, W. B., Jr., Gómez, Á., Seyle, C., Morales, F., & Huici, C., 'Identity Fusion: The Interplay of Personal and Social Identities in Extreme Group Behaviour', *Journal of Personality and Social Psychology*, Vol. 96, No. 5, pp. 995–1011 (2009).

17 Whitehouse, Harvey, *Modes of Religiosity: A Cognitive Theory of Religious Transmission*, AltaMira Press (2004).

18 Whitehouse, H. & Lanman, J. A., 'The Ties That Bind Us', *Current Anthropology*, Vol. 55, No. 6, pp. 674 –95 (2014).

19 Whitehouse, Harvey, *Inside the Cult: Religious Innovation and Transmission in Papua New Guinea*, Oxford University Press (1995).

20 Richert, R. A., Whitehouse, H., & Stewart, E., 'Memory and Analogical Thinking in High-Arousal Rituals', in Whitehouse, H. & McCauley, R. N. (Eds.), *Mind and Religion: Psychological and Cognitive Foundations of Religiosity*, AltaMira Press (2005).

21 Pfeiffer, J. E., *The Creative Explosion: An Inquiry into the Origins of Art and Religion*, Cornell University Press (1985).

22 Sonic Arts Research Centre website, Queen's University Belfast. Available at: www.qub.ac.uk/sarc/

23 Jong, J., Whitehouse, H., Kavanagh, C., & Lane, J., 'Shared Negative Experiences Lead to Identity Fusion via Personal Reflection', *PLoS ONE*, Vol. 10, No. 12, e0145611 (2015). DOI: 10.1371/journal.pone.0145611

24 Whitehouse, H., et al., 'The Evolution of Extreme Cooperation via Shared Dysphoric Experiences'.

25 Newson, M., Buhrmester, M., & Whitehouse, H., 'United in Defeat: Shared Suffering and Group Bonding Among Football Fans', *Managing Sport and Leisure*, Vol. 28, No. 2, pp. 164–81 (2021).

26 Kavanagh, C. M., Kapitány, R., Putra, I. E., & Whitehouse, H., 'Exploring the Pathways Between Transformative Group Experiences and Identity Fusion', *Frontiers in Psychology*, Vol. 11, No. 1172 (2020). DOI: 10.3389/fpsyg.2020.01172

27 Buhrmester, M., Burnham, D., Johnson, D., Curry, O.S., Macdonald, D., & Whitehouse, H., 'How Moments Become Movements: Shared Outrage, Group Cohesion, and the Lion That Went Viral', *Frontiers*, Vol. 6 (2018). DOI: 10.3389/fevo.2018.00054
28 Tasuji, T., Reese, E., van Mulukom, V., & Whitehouse, H., 'Band of Mothers: Childbirth as a Female Bonding Experience', *PLoS ONE*, Vol. 15, No. 10, e0240175 (2020). DOI: 10.1371/journal.pone.0240175
29 Swann, W. B., Jr., et al., 'Identity Fusion: The Interplay of Personal and Social Identities in Extreme Group Behavior'.
30 Swann, W. B., Jr., Gómez, Á., Huici, C., Morales, J. F., & Hixon, J. G., 'Identity Fusion and Self-Sacrifice: Arousal as a Catalyst of Pro-group Fighting, Dying, and Helping Behavior', *Journal of Personality and Social Psychology*, Vol. 99, No. 5, pp. 824–41 (2010).
31 Swann, W. B., Jr., et al., 'Dying and Killing for One's Group'.
32 Whitehouse, H., McQuinn, B., Buhrmester, M. D., & Swann, W. B., Jr., 'Brothers in Arms: Warriors Bond Like Family', *Proceedings of the National Academy of Sciences USA (PNAS)*, Vol. 111, No. 50, pp. 17783–5 (2014).
33 Whitehouse, H., et al., 'The Evolution of Extreme Cooperation via Shared Dysphoric Experiences'; Buhrmester, M. D., Newson, M., Vázquez, A., Wallisen, T. H., & Whitehouse, H., 'Winning at Any Cost: Identity Fusion, Group Essence, and Maximizing Ingroup Advantage, *Self and Identity*, Vol. 17, No. 5, pp. 500–516 (2018).
34 Vázquez, A., Ordoñana, J. R., Whitehouse, H., & Gómez, Á., 'Why Die for My Sibling? The Positive Association Between Identity Fusion and Imagined Loss with Endorsement of Self-sacrifice / ¿Por qué morir por un hermano? La asociación positiva entre la fusión de identidad y la pérdida imaginada con la disposición al autosacrificio', *Revista de Psicología Social*, Vol. 34, No. 3, pp. 413–38 (2019).
35 Gómez, Á., Bélanger, J. J., Chinchilla, J., et al., 'Admiration for Islamist Groups Encourages Self-sacrifice Through Identity Fusion', *Humanities and Social Sciences Communications*, Vol. 8, No. 54 (2021). DOI: 10.1057/s41599-021-00734-9
36 https://en.wikipedia.org/wiki/John_R._Fox#cite_note-:12-6

4. Cranking Up Conformism

1 Boyd, R., Schonmann, R. H., & Vicente, R., 'Hunter-Gatherer Population Structure and the Evolution of Contingent Cooperation', *Evolution and Human Behavior*, Vol. 35, No. 3, pp. 219–27 (2014).

2 Dietrich, O., Notroff, J., & Schmidt, K., 'Feasting, Social Complexity, and the Emergence of the Early Neolithic of Upper Mesopotamia: A View from Göbekli Tepe', in Chacon, R. & Mendoza, R. (Eds.), *Feast, Famine or Fighting? Studies in Human Ecology and Adaptation*, Vol. 8, Springer (2017).

3 Borrell, F. & Molist, M., 'Projectile Points, Sickle Blades and Glossed Points: Tools and Hafting Systems at Tell Halula (Syria) During the 8th Millennium cal. BC', in *Paléorient*, Vol. 33, No. 2, pp. 59–77 (2007).

4 Akkermans, Peter M. M. G. & Schwartz, Glenn M., *The Archaeology of Syria: From Complex Hunter-Gatherers to Early Urban Societies (c. 16,000–300 BC)*, Cambridge University Press (2004).

5 Mithen, S., 'From Ohalo to Çatalhöyük: The Development of Religiosity During the Early Prehistory of Western Asia, 20,000–7000 BC', in Whitehouse, H. & Martin, L. H. (Eds.), *Theorizing Religions Past: Historical and Archaeological Perspectives*, AltaMira Press (2004).

6 Whitehouse, Harvey, *Inside the Cult: Religious Innovation and Transmission in Papua New Guinea*, Oxford University Press (1995).

7 Whitehouse, Harvey, *Arguments and Icons: Divergent Modes of Religiosity*, Oxford University Press (2000).

8 Whitehouse, Harvey, 'Memorable Religions: Transmission, Codification, and Change in Divergent Melanesian Contexts', *Man* (New Series), Vol. 27, No. 4, pp. 777–97 (1992).

9 Barth, Fredrik, *Cosmologies in the Making: A Generative Approach to Cultural Variation in Inner New Guinea*, Cambridge University Press (1987).

10 For a much more detailed discussion of both of these points and of the psychology involved, see Whitehouse, Harvey, *Arguments and Icons: Divergent Modes of Religiosity;* Whitehouse, Harvey, *Modes of Religiosity: A Cognitive Theory of Religious Transmission*, AltaMira Press (2004); and Whitehouse, Harvey, *The Ritual Animal,* Oxford University Press (2021).

11 Anderson, Benedict, *Imagined Communities: Reflections on the Origin and Spread of Nationalism*, Verso Books (1991).

12 Whitehouse, H., 'Appropriated and Monolithic Christianity in Melanesia', in Cannell, F. (Ed.), *The Anthropology of Christianity*, Duke University Press (2006).

13 Whitehouse, H. & Hodder, I., 'Modes of Religiosity at Çatalhöyük', in Hodder, I. (Ed.), *Religion in the Emergence of Civilization: Çatalhöyük as a Case Study*, Cambridge University Press (2010); Whitehouse, H., Mazzucato, C., Hodder, I., & Atkinson, Q. D., 'Modes of Religiosity and the Evolution of Social Complexity at Çatalhöyük', in Hodder, I. (Ed.), *Religion at Work in a Neolithic Society: Vital Matters*, Cambridge University Press (2014).

14 Hodder, Ian & Tsoraki, Christina (Eds.), *Communities at Work: The Making of Çatalhöyük*, Çatalhöyük Research Project Series 15, British Institute at Ankara (2022).

15 Excerpt from 'The Story of God with Morgan Freeman: Interview with Harvey Whitehouse'. Available at: www.youtube.com/watch?v=IZhJicFWEuo

16 Atkinson, Q. D. & Whitehouse, H., 'The Cultural Morphospace of Ritual Form', *Evolution and Human Behaviour*, Vol. 32, No. 1, pp. 50–62 (2011).

17 Ember, C. R. & Ember, M., 'Cross-cultural Research', in Bernard, H. R. (Ed.), *Handbook of Methods in Cultural Anthropology*, AltaMira (1998).

18 Murdock, George Peter, *Ethnographic Atlas: A Summary*, University of Pittsburgh Press (1967).

19 Gantley, M., Whitehouse, H., & Bogaard, A., 'Material Correlates Analysis (MCA): An Innovative Way of Examining Questions in Archaeology Using Ethnographic Data', *Advances in Archaeological Practice*, Vol. 6, No. 4, pp. 328–41 (2018).

20 Ibid.

21 Sahlins, Marshall, *Stone Age Economics*, Tavistock (1974).

22 Byrd, B. F., 'Public and Private, Domestic and Corporate: The Emergence of the Southwest Asian Village', *American Antiquity*, Vol. 59, No. 4, pp. 639–66 (1994); Byrd, B. F., 'Reassessing the Emergence of Village Life in the Near East', *Journal of Archaeological Research*, Vol. 13, No. 3, pp. 231–89 (2005).

23 Kuijt, I., Guerrero, E., Molist, M., & Anfruns, J., 'The Changing Neolithic Household: Household Autonomy and Social Segmentation,

Tell Halula, Syria', *Journal of Anthropological Archaeology*, Vol. 30, No. 4, pp. 502–22 (2011).

24 Düring, B. & Marciniak, A., 'Households and Communities in the Central Anatolian Neolithic', *Archaeological Dialogues*, Vol. 12, No. 2, pp. 165–87 (2006); Marciniak, A., 'Communities, Households and Animals: Convergent Developments in Central Anatolian and Central European Neolithic', *Documenta Praehistorica*, Vol. 35 pp. 93–109 (2008).

25 Hodder, I., 'The Vitalities of Çatalhöyük', in Hodder, I. (Ed.), *Religion at Work in a Neolithic Society: Vital Matters*, Cambridge University Press (2014).

26 Hodder, I. & Pels, P., 'History Houses: A New Interpretation of Architectural Elaboration at Çatalhöyük', in Hodder, I. (Ed.), *Religion in the Emergence of Civilization*, Cambridge University Press (2010).

27 Baird, D., Fairbairn, A., & Martin, L., 'The Animate House, the Institutionalization of the Household in Neolithic Central Anatolia', *World Archaeology*, Vol. 49, No. 5, pp. 753–76 (2016).

28 Kuijt, I., 'Material Geographies of House Societies: Reconsidering Neolithic Çatalhöyük, Turkey', *Cambridge Archaeological Journal*, Vol. 28, No. 4, pp. 565–90 (2018).

29 Meyer, Fortes, *The Dynamics of Clanship Among the Tallensi*, Oxford University Press (1945).

30 Malinowski, Bronislaw, *Argonauts of the Western Pacific*, Routledge (2014).

5. Religiosity and the Rise of Supernatural Authority

1 Raddato, C., 'Minoan Storage Jars at the Palace of Knossos', *World History Encyclopedia* (2019). Available at: www.worldhistory.org/image/10598/minoan-storage-jars-at-the-palace-of-knossos/

2 Molloy, B. P. C., 'Martial Minoans? War as Social Process, Practice and Event in Bronze Age Crete', *The Annual of the British School at Athens*, Vol. 107, pp. 87–142 (2012). Available at: www.jstor.org/stable/41721880

3 Wall, S. M., Musgrave, J. H., & Warren, P. M., 'Human Bones from a Late Minoan IB House at Knossos', *The Annual of the British School at Athens*, Vol. 81, pp. 333–88 (1986).

4 Kuijt, Ian (Ed.), *Life in Neolithic Farming Communities*, Kluwer Academic/Plenum Publishers (2000).

5 Hertz, R., 'A Contribution to a Study of the Collective Representation of Death', in Mauss, M., Hubert, H., & Hertz, R. (Eds.), *Saints, Heroes, Myths, and Rites: Classical Durkheimian Studies of Religion and Society*, Routledge (2009).

6 Peoples, H. C., Duda, P., & Marlowe, F. W., 'Hunter-Gatherers and the Origins of Religion', *Human Nature*, Vol. 27, pp. 261–82 (2016).

7 Sheils, D., 'Toward a Unified Theory of Ancestor Worship: A Cross-cultural Study', *Social Forces*, Vol. 54, No. 2, pp. 427–40 (1975).

8 Sahlins, M., 'Poor Man, Rich Man, Big-Man, Chief: Political Types in Melanesia and Polynesia', *Comparative Studies in Society and History*, Vol. 5, No. 3, pp. 285–303 (1963).

9 Godelier, M. & Strathern, M. (Eds.), *Big Men and Great Men: Personifications of Power in Melanesia*, Cambridge University Press (1991).

10 Trigger, Bruce G., *Understanding Early Civilizations*, Cambridge University Press (2003).

11 Postgate, J. N., *Early Mesopotamia: Society and Economy at the Dawn of History*, Routledge (1992).

12 Steinkeller, Piotr, 'The Divine Rulers of Akkade and Ur: Toward a Definition of the Deification of Kings in Babylonia', in *History, Texts and Art in Early Babylonia: Three Essays*, De Gruyter (2017).

13 Friedman, J., 'Tribes, States, and Transformations', in Bloch, M. (Ed.), *Marxist Analyses and Social Anthropology*, Routledge (1984).

14 Firth, Raymond, *Rank and Religion in Tikopia: A Study in Polynesian Paganism and Conversion to Christianity*, Allen & Unwin (1970).

15 Whitehouse, H., 'From Possession to Apotheosis: Transformation and Disguise in the Leadership of a Cargo Movement', in Feinberg, R. & Watson-Gegeo, K. A. (Eds.), *Leadership and Change in the Western Pacific*, Athlone (1996).

16 Liu, Yue-Chen, Hunter-Anderson, R., Cheronet, O., et al., 'Ancient DNA Reveals Five Streams of Migration into Micronesia and Matrilocality in Early Pacific Seafarers', *Science*, Vol. 377, No. 6601, pp. 72–9 (2022).

17 Boone, E. H. (Ed.), *Ritual Human Sacrifice in Mesoamerica*, Dumbarton Oaks (1984).

18 Ceruti, M. C., 'Frozen Mummies from Andean Mountaintop Shrines: Bioarchaeology and Ethnohistory of Inca Human Sacrifice', *BioMed Research International*, Vol. 2015, No. 439428 (2015). DOI: 10.1155/2015/439428

19 Watts, J., Sheehan, O., Atkinson, Q. D., et al., 'Ritual Human Sacrifice Promoted and Sustained the Evolution of Stratified Societies', *Nature*, Vol. 532, pp. 228–31 (2016).

20 Burke, B. L., Martens, A., & Faucher, E. H., 'Two Decades of Terror Management Theory: A Meta-analysis of Mortality Salience Research', *Personality and Social Psychology Review*, Vol. 14, pp. 155–95 (2010); Greenberg, J., Solomon, S., & Arndt, J., 'A Basic but Uniquely Human Motivation: Terror Management', in Shah, J. Y. & Gardner, W. L. (Eds.), *Handbook of Motivation Science*, The Guilford Press (2008).

21 Bloch, Maurice, *Placing the Dead: Tombs, Ancestral Villages and Kinship Organization in Madagascar* (Studies in Anthropology), Academic Press (1971).

22 Covey, R. Alan, *Inca Apocalypse: The Spanish Conquest and the Transformation of the Andean World*, Oxford University Press (2020).

23 Turchin, P., Whitehouse, H., François, P., et al., 'An Introduction to Seshat: Global History Databank', *Journal of Cognitive Historiography*, Vol. 5, No. 1–2, pp. 115–23 (2020).

24 http://seshatdatabank.info/methods/world-sample-30/

25 Many of these had to be further subdivided, creating a much longer list of features. So, for example, among our fifty-one main variables were six features relating to systems of information storage. But when it came to the types of scripts being used, we broke this down further into a series of more fine-grained elements, each of which could be coded as present, absent, or unknown in the societies and periods of history we wanted to code, such as: lists, tables, classifications; calendar; sacred texts; religious literature; practical literature; history; philosophy; scientific literature; fiction (including poetry). So, although you could say that we had just fifty-one measures of social complexity, we actually had a lot more than that.

26 Turchin, P., Currie, T. E., Whitehouse, H., et al., 'Quantitative Historical Analysis Uncovers a Single Dimension of Complexity that Structures Global Variation in Human Social Organization', *Proceedings of the*

National Academy of Sciences USA (PNAS), Vol. 115, No. 2, E144–E151 (2018). DOI: 10.1073/pnas.1708800111

27 D'Altroy, Terence N., *The Incas: The Peoples of America* (Second Edition), Wiley Blackwell (2014).

28 Turchin, Peter, *Ultrasociety: How 10,000 Years of War Made Humans the Greatest Cooperators on Earth*, Beresta Books (2016).

29 Ibid.

30 Turchin, P., 'The Evolution of Moralizing Supernatural Punishment: Empirical Patterns', in Larson, J., Reddish, J., & Turchin, P. (Eds.), *The Seshat History of Moralizing Religion*, Beresta Books (in press).

31 Jaspers, Karl, *The Origin and Goal of History* (trans. M. Bullock), Yale University Press (1953); Eisenstadt, S. N., *Japanese Civilization: A Comparative View*, University of Chicago Press (1996); Bellah, Robert N., *Religion in Human Evolution: From the Paleolithic to the Axial Age*, Harvard University Press (2011).

32 Mullins, D. A., Hoyer, D., Collins, C., Currie, T., Feeney, K., François, P., Savage, P. E., Whitehouse, H., & Turchin, P., 'A Systematic Assessment of the Axial Age Thesis Using Global Comparative Historical Evidence', *American Sociological Review*, Vol. 83, No. 3, pp. 596–626 (2018).

33 Ebrey, Patricia, Walthall, Ann, & Palais, James, *East Asia: A Cultural, Social, and Political History*, Houghton Mifflin Harcourt (2006).

34 Whitehouse, H., François, P., Cioni, E., Levine, J., Hoyer, D., Reddish, J., & Turchin, P., 'Conclusion: Was There Ever an Axial Age?', in Hoyer, D. & Reddish, J. (Eds.), *The Seshat History of the Axial Age*, Beresta Books (2019).

35 Turchin, P., Whitehouse, H., Gavrilets, S., et al., 'Disentangling the Evolutionary Drivers of Social Complexity: A Comprehensive Test of Hypotheses', *Science Advances*, Vol. 8, No. 25 (2022). DOI: 10.1126/sciadv.abn3517

36 Alan Strathern describes these as 'transcendentalist' religions in his book *Unearthly Powers: Religious and Political Change in World History*, Cambridge University Press (2019).

37 Worsley, Peter, *The Trumpet Shall Sound: A Study of 'Cargo' Cults in Melanesia*, Schocken Books (1968).

6. Tribalism and the Evolution of Warfare

1 Wittfogel, Karl A., *Oriental Despotism: A Comparative Study of Total Power*, Oxford University Press (1957).
2 Fried, Morton H., *The Evolution of Political Society: An Essay in Political Anthropology*, Random House (1967).
3 Oppenheimer, Franz, *The State: Its History and Development Viewed Sociologically*, Free Life Editions (1975); Carneiro, R. L., 'A Theory of the Origin of the State', *Science*, Vol. 169, pp. 733–8 (1970).
4 Turchin, P., Whitehouse, H., Gavrilets, S., et al., 'Disentangling the Evolutionary Drivers of Social Complexity: A Comprehensive Test of Hypotheses', *Science Advances*, Vol. 8, No. 25 (2022). DOI: 10.1126/sciadv.abn3517
5 Turchin, Peter, *Ultrasociety: How 10,000 Years of War Made Humans the Greatest Cooperators on Earth*, Beresta Books (2016).
6 Algaze, G., 'Expansionary Dynamics of Some Early Pristine States', *American Anthropologist*, Vol. 95, pp. 304–33 (1993).
7 Guiart, J., 'Forerunners of Melanesian Nationalism', *Oceania*, Vol. 22, No. 2, pp. 81–90 (1951). Available at: www.jstor.org/stable/40328310
8 Barth, F., *Ritual and Knowledge Among the Baktaman of New Guinea*, Yale University Press (1975).
9 Schacter, D. L., Gilbert, D. T., & Wegner, D. M., 'Semantic and Episodic Memory', in *Psychology* (Second Edition), Worth (2009).
10 Tajfel, H. & Turner, J. C., 'The Social Identity Theory of Intergroup Behaviour', in Worchel, S. & Austin, W. G. (Eds.), *Psychology of Intergroup Relations*, Nelson-Hall (1986).
11 Whitehouse, H., 'Memorable Religions: Transmission, Codification, and Change in Divergent Melanesian Contexts', *Man* (New Series), Vol. 27, No. 4, pp. 777–97 (1992).
12 Turner, J. C. & Reynolds, K. J., 'The Story of Social Identity', in Postmes, T. & Branscombe, N. R. (Eds.), *Rediscovering Social Identity: Core Sources*, Psychology Press (2010).
13 Swann and Buhrmester describe the relationship between personal and groups identity as 'synergistic' in the case of fusion (the two working together) but 'hydraulic' in the case of identification (being

mutually repelling, a bit like oil and water). See: Swann, W. B., Jr. & Buhrmester, M. D., 'Identity Fusion', *Current Directions in Psychological Science*, Vol. 24, No. 1, pp. 52–7 (2015).

14 Social psychologist Marilyn Brewer questions the dominant view that identification automatically produces outgroup derogation but the research she cites in support of her view does not control for the effects of extended fusion, which might be predicted to reduce negative attitudes towards outgroups in the absence of competition or threat. See: Brewer, M. B., 'The Psychology of Prejudice: Ingroup Love and Outgroup Hate?', *Journal of Social Issues*, Vol. 55, No. 3, pp. 429–44 (1999).

15 Diener, E., Lusk, R., Defour, D., & Flax, R., 'Deindividuation: Effects of Group Size, Density, Number of Observers, and Group Member Similarity on Self-consciousness and Disinhibited Behavior', *Journal of Personality and Social Psychology*, Vol. 39, No. 3, pp. 449–59 (1980).

16 Swann, W. B., Jr., Gómez, Á., Seyle, C., & Morales, F., 'Identity Fusion: The Interplay of Personal and Social Identities in Extreme Group Behavior', *Journal of Personality and Social Psychology*, Vol. 96, No. 5, pp. 995–1011 (2009).

17 Johnson, R. D. & Downing, L. L., 'Deindividuation and Valence of Cues: Effects on Prosocial and Antisocial Behavior', *Journal of Personality and Social Psychology*, Vol. 37, No. 9, pp. 1532–8 (1979).

18 Besta, T., Gómez, Á., & Vázquez, A., 'Readiness to Deny Group's Wrongdoing and Willingness to Fight for Its Members: The Role of Poles' Identity Fusion with the Country and Religious Group', *Current Issues in Personality Psychology*, Vol. 2, No. 1, pp. 49–55 (2014); Fredman, L. A., Buhrmester, M. D., Gómez, Á., Fraser, W. T., Talaifar, S., Brannon, S. M., & Swann, W. B., Jr., 'Identity Fusion, Extreme Progroup Behavior, and the Path to Defusion', *Social and Personality Psychology Compass*, Vol. 9, No. 9, pp. 468–80 (2015); Buhrmester, M. D. & Swann, W. B., Jr., 'Identity Fusion', in Scott, R. A. & Kosslyn, S. M. (Eds.), *Emerging Trends in the Social and Behavioral Sciences*, John Wiley (2015); Swann, W. B., Jr., et al., 'Identity Fusion: The Interplay of Personal and Social Identities in Extreme Group Behavior'.

19 Swann, W. B., Jr., Gómez, Á., Buhrmester, M. D., López-Rodríguez, L., Jiménez, J., & Vázquez, A., 'Contemplating the Ultimate Sacrifice:

Identity Fusion Channels Pro-group Affect, Cognition, and Moral Decision-Making', *Journal of Personality and Social Psychology*, Vol. 106, No. 5, pp. 713–27 (2014); Swann, W. B., Jr., Gómez, Á., Dovidio, J. F., Hart, S., & Jetten, J., 'Dying and Killing for One's Group: Identity Fusion Moderates Responses to Intergroup Versions of the Trolley Problem', *Psychological Science*, Vol. 21, No. 8, pp. 1176–83 (2010).

20 Van Mulukom, V., Debeuf, K., Atalay, E. D., & Whitehouse, H., 'What Makes Muslim Minorities Willing to Fight and Die? Exploring the Role of Threat and Group Psychology' (submitted).

21 White, F. A., Newson, M., Verrelli, S., & Whitehouse, H., 'Pathways to Prejudice and Outgroup Hostility: Group Alignment and Intergroup Conflict Among Football Fans', *Journal of Applied Social Psychology*, Vol. 51, No. 7 (2021). DOI: 10.1111/jasp.12773

22 Buhrmester, M. D., Fraser, W. T., Lanman, J. A., Whitehouse, H., & Swann, W. B., Jr., 'When Terror Hits Home: Identity Fused Americans Who Saw Boston Bombing Victims as "Family" Provided Aid', *Self & Identity*, Vol. 14, No. 3, pp. 253–70 (2015).

23 For a summary of all this research see Whitehouse, Harvey, *The Ritual Animal: Imitation and Cohesion in the Evolution of Social Complexity*, Oxford University Press (2021).

24 Hunter, L. W. & Handford, S. A., *Aineiou Poliorketika: Aeneas on Siegecraft*, Clarendon Press (1927).

25 Schofield, A., 'Keeping It Together: Aeneas Tacticus and Unit Cohesion in Ancient Greek Siege Warfare', in Hall, J. R., Rawlings, L., & Lee, G. (Eds.), *Unit Cohesion and Warfare in the Ancient World: Military and Social Approaches*, Routledge (2023).

26 Hall, J. R., 'Unit Cohesion in the Ancient World: An Introduction', in Hall, J. R., et al., *Unit Cohesion and Warfare in the Ancient World*.

27 Konijnendijk, R., 'The Eager Amateur: Unit Cohesion and the Athenian Hoplite Phalanx', in Hall, J. R., et al., *Unit Cohesion and Warfare in the Ancient World*.

28 Walbank, F. W., *A Historical Commentary on Polybius*, Vol. 1, Oxford University Press (1957).

29 Mufti, M., 'Jihad as Statecraft: Ibn Khaldun on the Conduct of War and Empire', *History of Political Thought*, Vol. 30, No. 3, pp. 385–410 (2009).

30 Barstad, Hans M., *History and the Hebrew Bible: Studies in Ancient Israelite and Ancient Near Eastern Historiography*, Mohr Siebeck (2008).
31 Atran, Scott, *Talking to the Enemy: Violent Extremism, Sacred Values, and What It Means to Be Human*, Penguin (2011).
32 Atran, Scott, 'The Will to Fight', *Aeon* (2022). Available from: https://aeon.co/essays/wars-are-won-by-people-willing-to-fight-for-comrade-and-cause
33 Tossell, C. C., Gómez, Á., Visser, E. J., Vázquez, A., Donadio, B. T., Metcalfe, A., Rogan, C., Davis, R., & Atran, S., 'Spiritual Over Physical Formidability Determines Willingness to Fight and Sacrifice Through Loyalty in Cross-cultural Populations', *Proceedings of the National Academy of Sciences USA (PNAS)*, Vol. 119, No. 6, e2113076119 (2022). DOI: 10.1073/pnas.2113076119

7. Conformism and the Climate

1 UN Environmental Programme, 'Emissions Gap Report' (27 October 2022). Available at: www.unep.org/resources/emissions-gap-report-2022
2 HDI Global, 'The Future of Food: What Will You Be Eating in 2050?' (14 October 2021). Available at: www.hdi.global/infocenter/insights/2021/future-of-food/
3 Reinhardt, L. & Whitehouse, H., 'What Kinds of Speeches Motivate Climate Action?' (in preparation).
4 Whitehouse, Harvey, *Inside the Cult: Religious Innovation and Transmission in Papua New Guinea*, Oxford University Press (1995). The quoted material comes from page 1.
5 Ligaiula, P., 'Port Vila Call for a Just Transition to a Fossil Fuel Free Pacific', Pacific News Service (17 March 2023).
6 Liu, Yue-Chen, Hunter-Anderson, R., Cheronet, O., et al., 'Ancient DNA Reveals Five Streams of Migration into Micronesia and Matrilocality in Early Pacific Seafarers', *Science*, Vol. 377, No. 6601, pp. 72–9 (2022).
7 Stebbins, T. N., 'Mali Baining Perspectives on Language and Culture Stress', *International Journal of the Sociology of Language*, Vol. 2004, No. 169, pp. 161–75 (2004).

8 Weiner, Annette B., *Women of Value, Men of Renown: New Perspectives in Trobriand Exchange*, University of Texas Press (1983).
9 Mackey, John E. & Sisodia, Raj, *Conscious Capitalism: Liberating the Heroic Spirit of Business*, Harvard Business Review Press (2012).
10 Raworth, Kate, *Doughnut Economics: Seven Ways to Think Like a 21st-Century Economist*, Penguin Random House (2017).
11 Caşu, D., 'Attempts at Introducing Communist Rituals in Family Traditions and Holidays: Case Study on Moldavian SSR', *Codrul Cosminului*, Vol. 20, No. 1, pp. 273–8 (2014).
12 Johnson, I., '"Ruling Through Ritual": An Interview with Guo Yuhua', *The China File* (2018): www.chinafile.com/library/nyrb-china-archive/ruling-through-ritual-interview-guo-yuhua
13 Rybanska, V., McKay, R., Jong, J., & Whitehouse, H., 'Rituals Improve Children's Ability to Delay Gratification', *Child Development*, Vol. 89, No. 2, pp. 349–59 (2018).
14 Pew Research Center, 'Religious Landscape Study' (2014). Available at: www.pewresearch.org/religion/about-the-religious-landscape-study/
15 Pew Research Center, 'Young Adults Around the World Are Less Religious by Several Measures' (2018). Available at: www.pewresearch.org/religion/2018/06/13/young-adults-around-the-world-are-less-religious-by-several-measures/
16 Torchalla, I., Li, K., Strehlau, V., Linden, I. A., & Krausz, M., 'Religious Participation and Substance Use Behaviors in a Canadian Sample of Homeless People', *Community Mental Health Journal*, Vol. 50, No. 7, pp. 862–9 (2014).
17 Putnam, Robert D. & Campbell, David E., *American Grace: How Religion Divides and Unites Us*, Simon and Schuster (2010).
18 Chater, N. & Loewenstein, G., 'The I-frame and the S-frame: How Focusing on Individual-Level Solutions Has Led Behavioral Public Policy Astray', *Behavioral and Brain Sciences*, Vol. 46, No. E147 (2023). DOI: 10.1017/S0140525X22002023
19 Beermann, V., Rieder, A., & Uebernickel, F., 'Green Nudges: How to Induce Pro-environmental Behavior Using Technology', International Conference on Information Systems (2022).
20 https://citizensassembly.co.uk

8. Religiosity on Sale

1 New Creation Church may be exceptionally tolerant in this regard – not all forms of evangelical Christianity would be so accepting of members subscribing to multiple faith traditions at once.

2 Zengkun, F., 'Singapore Pastor Joseph Prince Goes Worldwide', *Straits Times* (28 October 2014).

3 It should be noted that different evangelical churches offer different kinds of practical benefits – including insurance schemes in some cases (see Auriol, E., Lassebie, J., Panin, A., Raiber, E., & Seabright, P., 'God Insures Those Who Pay? Formal Insurance and Religious Offerings in Ghana', *Quarterly Journal of Economics*, Vol. 135, No. 4, pp. 1799–1848 (2020)).

4 'Tithes and Offerings', New Creation Church website: www.newcreation.org.sg/give/

5 Iannaccone, L. R., 'Sacrifice and Stigma: Reducing Free-Riding in Cults, Communes, and Other Collectives', *Journal of Political Economy*, Vol. 100, No. 2, pp. 271–91 (1992).

6 For a discussion of how religions can signal mate value in sexual selection, see Slone, J. & Van Slyke, J. (Eds.), *The Attraction of Religion: A New Evolutionary Psychology of Religion*, Bloomsbury Academic (2015).

7 Seabright, Paul, *The Divine Economy: How Religions Compete for Wealth, Power and People*, Princeton University Press (in press).

8 European Research Council, 'Unboxing Cultural Rituals: Christmas in Pandemic Times' (2020): https://erc.europa.eu/projects-figures/stories/unboxing-cultural-rituals-christmas-pandemic-times

9 Pearson, B., 'Holiday Spending to Exceed $1 Trillion – And 11 Other Surprising Data Points of Christmas', *Forbes* (22 December 2016): www.forbes.com/sites/bryanpearson/2016/12/22/holiday-spending-to-exceed-1-trillion-and-11-other-surprising-data-points-of-christmas/?sh=1ef611e1247f

10 Although not 'supernatural' when manifested in real life as dementia or brain damage, such abnormal psychology is minimally counterintuitive in the sense of violating our intuitive expectations about mentalizing, and therefore intrinsically attention-grabbing, memorable, and transmittable according to MCI theory.

11 Tang, F., 'China's Lunar New Year Spending Growth Slows to Decade Low Despite Record US$148.96 Billion Sales', *South China Morning Post* (11 February 2019).

12 Guthrie, Stewart, *Faces in the Clouds: A New Theory of Religion*, Oxford University Press (1993).

13 Goffman, Erving, *The Presentation of Self in Everyday Life*, Doubleday Anchor (1959).

14 Guthrie, S., 'Bottles Are Men, Glasses Are Women: Religion, Gender, and Secular Objects', *Material Religion*, Vol. 3, No. 1, pp. 14–33 (2007).

15 Sharma, M. & Zillur, R., 'Anthropomorphic Brand Management: An Integrated Review and Research Agenda', *Journal of Business Research*, Vol. 149, pp. 463–75 (2022).

16 Strohminger, N. & Jordan, M. R., 'Corporate Insecthood', *Cognition*, Vol. 224, No. 105068 (2022). DOI: 10.1016/j.cognition.2022.105068

17 Ibid.

18 Moller, J. & Herm, S., 'Shaping Retail Brand Personality Perceptions by Bodily Experiences', *Journal of Retailing*, Vol. 89, No. 4, pp. 438–46 (2013).

19 Perez-Vega, R., Taheri, B., Farrington, T., & O'Gorman, K., 'On Being Attractive, Social and Visually Appealing in Social Media: The Effects of Anthropomorphic Tourism Brands on Facebook Fan Pages', *Tourism Management*, Vol. 66, pp. 339–47 (2018).

20 Huang, X., Li, X., & Zhang, M., '"Seeing" the Social Roles of Brands: How Physical Positioning Influences Brand Evaluation', *Journal of Consumer Psychology*, Vol. 23, No. 4, pp. 509–14 (2013).

21 Folse, J. A. G., Netemeyer, R. G., & Burton, S., 'Spokescharacters', *Journal of Advertising*, Vol. 41, No. 1, pp. 17–32 (2012).

22 Kucuk, S. U., 'Reverse (Brand) Anthropomorphism: The Case of Brand Hitlerization', *Journal of Consumer Marketing*, Vol. 37, No. 6, pp. 651–9 (2020).

23 I derive the following observations from the episode 'I Can't Believe It's Pink Margarine', *99% Invisible* podcast (19 October 2021): https://99percentinvisible.org/episode/i-cant-believe-its-pink-margarine/transcript/

24 See also 'Butter vs Margarine: One of America's Most Bizarre Food

Battles', *Distillations* podcast (15 November 2017): www.sciencehistory.org/stories/distillations-pod/butter-vs-margarine/

25 Slone, Jason D., *Theological Incorrectness: Why Religious People Believe What They Shouldn't,* Oxford University Press (2004).

26 www.buycott.com

27 https://donegood.co; https://blog.orangeharp.com.

28 Harcup, T. & O'Neill, D., 'What Is News? Galtung and Ruge Revisited', *Journalism Studies*, Vol. 2, No. 2, pp. 261–80 (2001).

29 www.niemanlab.org/2019/08/americans-with-less-education-are-more-likely-to-say-that-local-news-is-important-to-them-and-to-get-it-from-tv/ (2019); www.latimes.com/entertainment/envelope/la-xpm-2013-apr-02-la-et-ct-nielsen-educated-viewers-watch-less-tv-20130401-story.html (2013).

30 Fehr, E. & Fischbacher, U., 'Third-Party Punishment and Social Norms', *Evolution and Human Behavior*, Vol. 25, No. 2, pp. 63–87 (2004).

31 Geraci, A. & Surian, L., 'Preverbal Infants' Reactions to Third-Party Punishments and Rewards Delivered Toward Fair and Unfair Agents', *Journal of Experimental Child Psychology*, Vol. 226, No. 105574 (2023). DOI: 10.1016/j.jecp.2022.105574

32 Lee, Richard Borshay, *The !Kung San: Men, Women and Work in a Foraging Society*, Cambridge University Press (1979).

33 Morese, R., Rabellino, D., Sambataro, F., Perussia, F., Valentini, M. C., Bara, B. G., et al., 'Group Membership Modulates the Neural Circuitry Underlying Third Party Punishment', *PLoS ONE*, Vol. 11, No. 11, e0166357 (2016). DOI: 10.1371/journal.pone.0166357

34 Buckholtz, J. W., Asplund, C. L., Dux, P. E., Zald, D. H., Gore, J. C., Jones, O. D., & Marois, R., 'The Neural Correlates of Third-Party Punishment', *Neuron*, Vol. 60, No. 5, pp. 930–40 (2008).

35 Apps, M. A. J., McKay, R., Azevedo, R. T., Whitehouse, H., & Tsakiris, M., 'Not On My Team: Medial Prefrontal Cortex Contributions to Ingroup Fusion and Fairness', *Brain and Behaviour*, Vol. 8, No. 8 (2018). DOI: 10.1002/brb3.1030

36 Kantrowitz, A., 'The Man Who Built the Retweet: "We Handed a Loaded Weapon to 4-Year-Olds"', *BuzzFeed News* (23 July 2019).

37 Brady, W. J., Wills, J. A., Jost, J. T., Tucker, J. A., & Van Bavel, J. J., 'Emotion Shapes the Diffusion of Moralized Content in Social Networks', *Proceedings of the National Academy of Sciences USA (PNAS)*, Vol. 114, pp. 7313–18 (2017); Crockett, M. J., 'Moral Outrage in the Digital Age', *Nature Human Behaviour*, Vol. 1, pp. 769–71 (2017); Burton, J. W., Cruz, N., & Hahn, U., 'Reconsidering Evidence of Moral Contagion in Online Social Networks', *Nature Human Behaviour*, Vol. 5, pp. 1629–35 (2021).

38 Egebark, J. & Ekström, M., 'Like What You Like or Like What Others Like? Conformity and Peer Effects on Facebook', Research Institute of Industrial Economics, IFN Working Paper No. 886 (2011).

39 Jędryczka, W., Misiak, M., & Whitehouse, H., 'Explaining Political Polarization Over Abortion: The Role of Moral Values Among Conservatives', *Social Psychology*, Vol. 54, No. 4 (2023). DOI: 10.1027/1864-9335/a000525

40 Curry, O. S., 'Morality as Cooperation: A Problem-Centred Approach', in Shackelford, T. K. & Hansen, R. D. (Eds.), *The Evolution of Morality*, Springer International Publishing (2016).

41 'Digital Propaganda or "Normal" Political Polarization? Case Study of Political Debate on Polish Twitter', Panoptykon Foundation: https://panoptykon.org/twitter-report

42 Palese, M., 'The Irish Abortion Referendum: How a Citizens' Assembly Helped to Break Years of Political Deadlock', Electoral Reform Society (29 May 2018): www.electoral-reform.org.uk/the-irish-abortion-referendum-how-a-citizens-assembly-helped-to-break-years-of-political-deadlock/

43 Kantrowitz, A., 'The Man Who Built the Retweet'.

44 www.newscientist.com/article/2217937-whatsapp-restrictions-slow-the-spread-of-fake-news-but-dont-stop-it/

9. Tribalism Today

1 O'Brien, N., 'Interview with a Former Terrorist: Nasir Abbas' Deradicalization Work in Indonesia', *CTC Sentinel*, Vol. 1, No. 12 (2008).

2 Letzing, J. & Berkley, A., 'Is the Internet Really More Effective at Radicalizing People than Older Media?', *The Digital Economy*, World Economic Forum (2021). Available at: www.weforum.org/agenda/2021/07/is-the-internet-really-more-effective-at-radicalizing-people-than-older-media/

3 'The Use of Social Media by United States Extremists', National Consortium for the Study of Terrorism and Responses to Terrorism (START) (2023). Available at: www.start.umd.edu/pubs/START_PIRUS_UseOfSocialMediaByUSExtremists_ResearchBrief_July2018.pdf

4 Kavanagh, C. M., Kapitány, R., Putra, I. E., & Whitehouse, H., 'Exploring the Pathways Between Transformative Group Experiences and Identity Fusion', *Frontiers in Psychology*, Vol. 11, No. 1172 (2020). DOI: 10.3389/fpsyg.2020.01172

5 Courchesne, L., Rasikh, B., McQuinn, B., & Buntain, C., 'Powered by Twitter? The Taliban's Takeover of Afghanistan', Centre for Artificial Intelligence, Data, and Conflict (2022). Available at: www.tracesofconflict.com/_files/ugd/17ec87_19ecafa8cf1046af8554251bce0aaf6f.pdf

6 Ebner, J., Kavanagh, C., & Whitehouse, H., 'Is There a Language of Terrorists? A Comparative Manifesto Analysis', *Studies in Conflict and Terrorism* (2022). DOI: 10.1080/1057610X.2022.2109244

7 Buhrmester, M. D., Fraser, W. T., Lanman, J. A., Whitehouse, H., & Swann, W. B., Jr., 'When Terror Hits Home: Identity Fused Americans Who Saw Boston Bombing Victims as "Family" Provided Aid', *Self & Identity*, Vol. 14, No. 3, pp. 253–70 (2015).

8 Marks, G. & Miller, N., 'Ten Years of Research on the False-Consensus Effect: An Empirical and Theoretical Review', *Psychological Bulletin*, Vol. 102, No. 1, pp. 72–90 (1987).

9 Klein, J. W., Bastian, B., Odjidja, E. N., Ayaluri, S. S., Kavanagh, C. M., & Whitehouse, H., 'Ingroup Commitment Can Foster Intergroup Trust' (in preparation).

10 Suedfeld, P., Tetlock, P. E., & Streufert, S., 'Conceptual/Integrative Complexity', in Smith, C. P., Atkinson, J. W., McClelland, D. C., & Veroff, J. (Eds.), *Motivation and Personality: Handbook of Thematic Content Analysis*, Cambridge University Press (1992).

11 Suedfeld, P. & Tetlock, P., 'Integrative Complexity of Communications in International Crises', *Journal of Conflict Resolution*, Vol. 21, No. 1 (1977), pp. 169–84.
12 Suedfeld, P. & Bluck, S., 'Changes in Integrative Complexity Prior to Surprise Attacks', *Journal of Conflict Resolution*, Vol. 32, No. 4 (1988). DOI: 10.1177/0022002788032004002
13 Pruitt, D. G. & Lewis, S. A., 'Development of Integrative Solutions in Bilateral Negotiation', *Journal of Personality and Social Psychology*, Vol. 31, No. 4, pp. 621–33 (1975).
14 Johnson, Dominic, *Overconfidence and War: The Havoc and Glory of Positive Illusions*, Harvard University Press (2004).
15 Suedfeld, P., 'The Cognitive Processing of Politics and Politicians: Archival Studies of Conceptual and Integrative Complexity', *Journal of Personality*, Vol. 78, No. 6, pp. 1669–1702 (2010).
16 Ebner, J. & Whitehouse, H., 'Identity and Extremism: Sorting Out the Causal Pathways to Radicalisation and Violent Self-sacrifice', in Busher, J., Malkki, L., & Marsden, S. (Eds.), *Routledge Handbook on Radicalisation and Countering Radicalisation*, Routledge (2023).
17 This project is supported by the Calleva Centre at Magdalen College, Oxford.
18 Villettaz, P., Gillieron, G., & Killias, M., 'The Effects on Re-offending of Custodial vs. Non-custodial Sanctions: An Updated Systematic Review of the State of Knowledge', *Campbell Systematic Reviews*, Vol. 11, No. 1, pp. 1–92 (2015).
19 Suzanne Reich, Senior Lecturer (Criminology and Criminal Justice) and Program Director (School of Law and Justice, University of Southern Queensland) – personal communication.
20 Dobbie, W. S., and Yang, C. S., 'The Economic Costs of Pretrial Detention', *Research Briefs in Economic Policy*, No. 283 (2022). Available at: www.cato.org/research-briefs-economic-policy/economic-costs-pretrial-detention
21 Whitehouse, H. & Fitzgerald, R., 'Fusion and Reform: The Potential for Identity Fusion to Reduce Recidivism and Improve Reintegration', *Anthropology in Action*, Vol. 27, No. 1, pp. 1–13 (2020).

22 Newson, M., Bortolini, T. S., da Silva, R., Buhrmester, M., & Whitehouse, H., 'Brazil's Football Warriors: Social Bonding and Inter-group Violence', *Evolution and Human Behavior*, Vol. 39, No. 6, pp. 675–83 (2018).
23 Newson, M. & Whitehouse, H., 'The Twinning Project: How Football, the Beautiful Game, Can Be Used to Reduce Reoffending', *Prison Service Journal*, Vol. 248, pp. 28–31 (2020).
24 Newson, M., Peitz, L., Cunliffe, J., & Whitehouse, H., 'Pathways to Fusion in Prison: The Effects Attachment and Personal Transformation on the Twinning Project' (in preparation).
25 Newson, M., Peitz, L., Cunliffe, J., & Whitehouse, H., 'Improving Prison Behaviour and Receiving Community Attitudes Through Football' (in preparation).
26 Reich, S., Buhrmester, M. B., & Whitehouse, H., 'Identity Fusion and Going Straight: How Disclosure of Life-Changing Experiences Can Impact Ethical Hiring Practices and Positive Reintegration Outcomes', *Journal of Business Ethics* (under review).

Epilogue: Rise of the Teratribe

1 'The History of The World: Every Year': www.youtube.com/watch?v=-6Wu0Q7x5D0
2 www.parliament.scot/about/how-parliament-works/sustainability
3 www.oxfam.org/en/press-releases/richest-1-bag-nearly-twice-much-wealth-rest-world-put-together-over-past-two-years.
4 For a brief discussion of this point, see this interview with actor Morgan Freeman in the National Geographic Channel documentary series about the origins of religion: www.youtube.com/watch?v=IZhJicFWEuo
5 Whitehouse, H., 'From Conflict to Covid: How Shared Experiences Shape Our World and How They Could Improve It', *New England Journal of Public Policy*, Vol. 33, No. 2 (2021). Available at: https://scholarworks.umb.edu/nejpp/vol33/iss2/7/
6 Curry, O. S., Hare, D., Hepburn, C., Johnson, D. D. P., Buhrmester, M. D., Whitehouse, H., & Macdonald, D. W., 'Cooperative Conservation:

Seven Ways to Save the World', *Conservation Science and Practice*, Vol. 2, No. 1 (2019). DOI: 10.1111/csp2.123

7 Buhrmester, M. D., Burnham, D., Johnson, D. D. P., Curry, O. S., Macdonald, D. W., & Whitehouse, H., 'How Moments Become Movements: Shared Outrage, Group Cohesion, and the Lion That Went Viral', *Frontiers in Ecology and Evolution*, Vol. 6, No. 54 (2018). DOI: 10.3389/fevo.2018.00054

8 Jacobs, A. J., 'The World's Largest Family Reunion . . . We're All Invited!', TEDActive (2014). Available at: www.ted.com/talks/a_j_jacobs_the_world_s_largest_family_reunion_ we_re_all_invited?language=en

9 Reinhardt, L. & Whitehouse, H., 'Why Care for Humanity?' (submitted).

10 Cowan, M., 'Inclusiveness, Foresight, and Decisiveness: The Practical Wisdom of Barrier-Crossing Leaders', *New England Journal of Public Policy*, Vol. 29, No. 1, Article 14 (2017).

11 Buhrmester, M. D., Cowan, M. A., & Whitehouse, H., 'What Motivates Barrier-Crossing Leadership?', *New England Journal of Public Policy*, Vol. 34, No. 2, Article 7 (2022). Available at: https://scholarworks.umb.edu/nejpp/vol34/iss2/7/

Index